Optimization Techniques and Associated Applications

Currently, the techniques of operation research are widely used in every aspect of day-to-day life. This book discusses a variety of problems that arise in various businesses and develops mathematical theories as well as technology answers to solve them from an industry perspective.

Optimization Techniques and Associated Applications incorporates cutting-edge methods for locating early workable answers to an optimization challenge and acts as a road map for creating a catastrophe management paradigm that is sustainable. This book offers numerical methods for resolving mathematical issues that can be found in numerous sectors and includes specific case studies of actual industrial optimization applications. The uncertainty that arises in various businesses is explored and new and recently developed techniques are discussed.

Because this book primarily focuses on operations research and solutions to the challenges across a variety of disciplines, the audience is expansive and can include professionals, students, and researchers from mathematics as well as engineers from industrial engineering, computer science, information technology, mechanical, civil, electrical, petroleum, chemical, aerospace, aviation, meteorology, disaster management, and other departments.

Engineering Mathematics and Operations Research

Tools, Techniques, Theory, and Applications Used in Manufacturing and Management Science

Series Editor: Aliakbar Montazer Haghighi

The main aim of this new book series is to publish books by qualified authors in the areas related to mathematics and its applications in engineering and operations research areas. Thus, the series will publish rigorous and quality books related to a variety of areas of mathematics and science related to engineering. All books will include some applications, tools, and theories as well as the latest research and development of mathematics and its applications in engineering, manufacturing, and management sciences areas. Such topics that will be covered in this series are: vector analysis, linear algebra, fuzzy mathematics, sequences, series, probability, statistics, reliability, numerical analysis, approximation, teaching of mathematics for engineers, complex analysis, ordinary differential equations, differential equations, partial differential equations, stochastic processes, random walk, point process, Brownian motions, birth and death processes, and queueing theory.

Books in this series will be of interest to undergraduate and graduate students as course textbooks as well as reference books.

Optimization Techniques and Associated Applications
Edited by Dr Sandeep Singh and Dr Sandeep Dalal

For more information about this series, please visit: www.routledge.com/Engineering-and-Operations-Research/book-series/CRCEMORTTTA

Optimization Techniques and Associated Applications

Edited by
Dr Sandeep Singh and Dr Sandeep Dalal

CRC Press
Taylor & Francis Group
Boca Raton London New York

CRC Press is an imprint of the
Taylor & Francis Group, an **informa** business

Dedication

*Dedicated
To God*

Contents

Preface

The book *Optimization Techniques and Associated Applications* is a comprehensive exploration of advanced methodologies in optimization theory and their practical applications across various domains. This book brings together a collection of chapters authored by experts and researchers who delve into the intricate world of optimization and its real-world implications.

In today's fast-paced and complex world, optimization techniques play a pivotal role in enhancing efficiency, minimizing costs, and maximizing resource utilization across diverse fields ranging from engineering and mathematics to economics and sustainability. The chapters presented in this book reflect the breadth and depth of optimization theory, offering insights into cutting-edge methodologies and their applications in solving complex problems.

The journey through this book begins with a meticulous examination of redundant machining systems, public electric vehicle charging stations, transshipment problems, and operational controllability of higher-order integrodifferential systems. Readers will be introduced to sophisticated optimization algorithms such as particle swarm optimization (PSO), artificial bee colony (ABC) algorithms, and the EJMD method for intuitionistic fuzzy transportation problems.

Furthermore, this book delves into the realm of mathematical modeling and analysis, presenting solutions to physical system problems using partial differential equations. From the utilization of complex fuzzy matrices in assessing the sustainability of women entrepreneurship to the application of discrete Fourier series using generalized difference operators, each chapter offers valuable insights into the multifaceted landscape of optimization.

Moreover, emerging trends in digital education and the utilization of soft trees for spanning applications underscore the dynamic nature of optimization techniques in addressing contemporary challenges. The exploration of fuzzy random multi-objective quadratic transportation problems and the review of optimization strategies inspired by nature highlight the interdisciplinary nature of optimization research.

As editors, we are deeply grateful to the contributors who have dedicated their expertise and knowledge to this collective endeavor. Their meticulous research and scholarly insights have enriched the content of this book, making it a valuable resource for academics, researchers, practitioners, and students alike.

We hope that *Optimization Techniques and Associated Applications* serves as a catalyst for innovation and inspires further exploration in the fascinating field of optimization. May the insights shared within these pages ignite new ideas, foster collaboration, and contribute to the advancement of knowledge and technology in pursuit of a more optimized world.

Editors

Dr. Sandeep Singh is an Assistant Professor with the Department of Mathematics at Akal University Talwandi Sabo, Bathinda, Punjab, India. He received his Ph.D. in Group Theory (Mathematics) from School of Mathematics, Thapar University, Patiala, India, and M.Sc. degree in Mathematics from Punjabi University, Patiala, India. He was also a postdoctoral fellow at the Department of Mathematics, Indian Institute of Technology Roorkee under program SERB-NPDF (National Postdoctoral Fellowship). His research interests include Group Theory, Automorphism Groups, Number Theory, Sumset Problems, and Optimization Techniques. Besides holding an excellent academic record throughout, he qualified the national level examination, NET-JRF, conducted by UGC-CSIR, India. Also, he had been a recipient of CSIR junior and senior research fellowship for 2011–2013 and 2013–2016, respectively. He has published more than 20 research papers in various journals of international repute and of different publishers like Springer, Taylor Francis, World Scientific, and others. He has presented his research work at various international conferences, and he is the recipient of travel support by the Department of Science and Technology (DST), India. He has a teaching experience of around 8 years. He has supervised more than three students for their M.Sc. dissertation work, and, currently, he has been supervising two Master's students at Akal University. He has also supervised two Ph.D. students, and, presently, two students are pursuing their Ph.D. under his supervision. He is a life member of Ramanujan Mathematical Society and Indian Mathematical Society and a reviewer of American Mathematical Society.

Dr. Sandeep Dalal received his M.Sc. degree in Mathematics from IIT Delhi, New Delhi. In 2016, he joined as a full-time Ph.D. scholar in the Department of Mathematics, Birla Institute of Technology and Science, Pilani, Pilani Campus, under the supervision of Dr. Jitender Kumar in the area of algebraic graph theory. He qualified the Graduate Aptitude Test for Engineering (GATE) for Mathematics in 2013 and the National Eligibility Test (NET) in 2017. He received a research excellence award from the Department of Mathematics, BITS Pilani, Pilani, in 2021. He is working in algebraic graph theory. He has published four papers in SCI indexed journals and two papers in Scopus indexed journals. He worked as an Assistant Professor at Akal University Talwandi Sabo, Bathinda, India, from October 2021 to January 2022. He taught real analysis, partial differential equation, and abstract algebra at UG level and integral equation and operation research at PG level. He was also a postdoctoral fellow at NISER, Bhubaneshwar, Orissa, India. Currently, he is working as an Assistant Professor at Punjab Engineering College, Chandigarh.

Contributors

P. Anukokila
Department of Mathematics
PSG College of Arts & Science
Coimbatore, India

Geeta Arora
Department of Mathematics
Lovely Professional University
Punjab, India

G. Dominic Babu
Department of Mathematics
Annaivelankanni College
Tholayavattam, Kanniyakumari

Kiran Bala
Department of Mathematics
Lovely Professional University
Punjab, India

Sahil Bansal
Department of Physics
Lajpat Rai DAV College
Jagraon, Punjab

D. Brightlin
Department of Mathematics
Annaivelankanni College
Tholayavattam, Kanniyakumari

Parmeet Kaur Chahal
Department of Mathematics
Central University of Jammu
Jammu, India

V. G. Deepa
Department of Mathematics
Sree Krishna College
Guruvayur, India

L. Jones Tarcius Doss
Department of Mathematics, College of
    Engineering
Anna University
Chennai, India

Bobin George
Department of Mathematics
Pavanatma College
Murickassery, India

Jinta Jose
Department of Science and Humanities
Viswajyothi College of Engineering and
    Technology
Vazhakulam, India

Karampreet Kaur
Department of Mathematics
Akal University
Talwandi Sabo, India

Kamlesh Kumar
Department of Mathematics
Central University of Jammu
Jammu, India

S. Aparna Lakshmanan
Department of Mathematics
Cochin University of Science and Technology
Cochin, India

K. T. Divya Mol
Department of Mathematics
St. Thomas College Palai
Kerala, India

B. Radhakrishnan
Department of Mathematics
PSG College of Technology
Coimbatore, India

Deepika Rani
Department of Mathematics
Dr. B. R. Ambedkar National Institute of
    Technology
Jalandhar, India

Lovelymol Sebastian
Department of Mathematics
St. Thomas College Palai
Kerala, India

Shalini Sharma
Department of Mathematics
Central University of Jammu
Jammu, India

Shivani
Department of Mathematics
Dr. B. R. Ambedkar National Institute of
    Technology
Jalandhar, India

V. N. Sreeja
Department of Mathematics
Sree Krishna College
Guruvayur, India

Rajesh K. Thumbakara
Department of Mathematics
Mar Athanasius College (Autonomous)
Kothamangalam, India

S. Yaazhini
Department of Mathematics, College of
    Engineering
Anna University
Chennai, India

1 Performance Analysis of Redundant Machining System under Generalized Triadic Policy

Parmeet Kaur Chahal, Kamlesh Kumar, and Shalini Sharma

1.1 INTRODUCTION

A *Queuing Model* is a feasible framework for expressing the service-focused problem in which customers arrive randomly in order to receive a service, with the service time being a random variable. In the context of queueing models for machining systems, customers refer to the failed machines that join the system to receive repair services. In queueing models of machining systems, we quite often see instances in which either the server is inactive or the failed machines waiting for service stack up due to the server's unavailability, resulting in a congestion condition in the system. To avoid such situations, the active number of servers present in the machining framework needs to get adjusted according to the built up queue size of machines that have failed in the system. Various control policies have been developed in the past in which different predefined thresholds are set to synchronize the number of active servers with the number of failed machines in the system. Yadin and Naor [17] were the ones who initially proposed one of these policies, which they referred to as N-policy. In accordance with N-policy, the server begins the service when the number of failed machines accumulated in the system reaches $N(\geq 1)$ and ends the service when the system turns into no failed machine available mode and the server remains inactive as soon as N-failed machines accumulate within the framework.

Bell [1] examined a control policy for optimizing a $M/M/2$ queueing model in which the count of two servers in active mode can be regulated on the basis of the queue size of the failed machines in the machining framework. Rhee and Sivazlian [11] introduced a very interesting and useful control policy and called it as *triadic*$(0, Q, N, M)$ *policy* for two server system and obtained the queue size distribution for $M/M/2$ queueing model. The definition of triadic policy $(0, Q, N, M)$ is outlined as follows: In the event the queueing system is empty, the two servers remain in dormant state until a certain preset threshold N for the count of failed machines is achieved. As soon as the count of failed machines attains the value, one among both the dormant servers will become active promptly. In cases where the count of failed machines arrives at a higher threshold $M(N \leq M)$, the other dormant server will be activated immediately. When both servers are active at the same time, and the count of failed machines falls to threshold $Q(1 < Q < N)$, then the server that ends up completing a service at that time will be deactivated. Furthermore, if one server is active and the count of failed machines in the repairable machining structure becomes zero, the server is deactivated immediately. The two servers will remain inactive until the above-mentioned thresholds are re-acquired. Wang and Chang [14] studied a two server queueing structure with redundancy performing in accordance with the triadic $(0, Q, N, M)$ policy. They determined that the reliability of the system can be improved by making use of standbys or additional servers. Wang and Wang [15] performed an optimal analysis for a finitely capacitated $M/M/2$ queueing model with triadic $(0, Q, N, M)$ policy. Ke et al. [5] further extended this model for infinite capacity queueing system with unreliable servers. Lin and Ke [10] employed the genetic algorithm technique for optimization

DOI: 10.1201/9781003407386-1

of multi-server infinitely capacitated queueing model under triadic policy. Haung et al. [4] made an attempt to control both arrivals and services for limited capacity $M/M/2$ queueing system operating according to the triadic control policy, and the authors optimize the cost function adopting the genetic algorithm method. Ketema et al. [6] in latest work examined a controllable $M/M/2$ repairable machining system along with working vacation and the triadic $(0, Q, N, M)$ policy. They established various system performance measures and deployed the grid search algorithm to optimize the cost function of the machining system.

The failure in various machining structures like the ones in the production system has an impact on the quality of the finished product and raises the overall cost of production. Therefore, we include the provision of standby machines or multi-servers to minimize these breakdowns in a machining system and allowing proper usage of given resources. Taylor and Jackson [13] considered the concept of standby machine in queueing modeling of machine repair problem using birth-death process for the very first time. The standby machines are categorized as hot, cold, and warm standbys. Warm standbys experience fewer failures than operating machines, whereas hot standbys experience the similar rate of failures as operating machines. The cold standby machines have zero failure rates. It is advantageous to use mixed standby machines rather than using single type of standbys due to some tech-no economic constraints. The cost difference of various types of standbys and the unavailability of required amount of particular type of standby favor the incorporation of mixed standbys to the machine repair system. A machining system with standbys provision is called a *redundant machining system*. Gnedenko et al. [3] investigated a machining system with numerous servers and mixed standbys. Kumar et al. [8] examined machine repair problems with ??-policy, two unreliable servers along with the provision of warm standbys. Gao and Wang [2] investigated a constant retrial redundant machine system having an unreliable service facility. The reliability and availability analysis was presented and thoroughly examined.

In many real-world congestion scenarios, the server either is unavailable to handle the failed machines or provides service with slower pace. The server may be unavailable for a variety of reasons, including routine maintenance, additional tasks, or simply taking a break. The server *vacation* is a period when a server remains momentarily inaccessible to execute repair services in the machining system. Levy and Yechiali [9] were the first to investigate the queueing framework combined with server vacations. The authors examined an extended queueing system in which the server is unavailable on account of a secondary job. In queueing modeling of machine repair problems, the working vacation mode for a server is considered when service is provided at a slower rate rather than completely shutting down the service. It is implied in order to avoid the loss caused by accumulated failed machines during the server's vacation period. In case the server heads back from a vacation and observes that the system is empty, then it goes for another vacation, and multiple vacation policy refers to this type of vacation situation. This concept of working vacation to the queueing modeling was introduced by Servi and Fin [12]. When working vacation is included in queueing models, the model becomes more realistic. Wu and Takagi [16] studied $M/G/1$ queueing system incorporating multiple working vacations for server and obtained various performance measures.

In realistic scenarios, it is observed that a redundant machining system cannot be limited to a problem involving only two servers. Consideration is given to the concept of multiple homogeneous or heterogeneous server problems. Kumar and Jain [7] investigated one such model having multiple heterogeneous servers which included redundancy and multiple working vacation policy. Using the SOR method, they calculated the system's steady-state probabilities. Taking this as motivation, we investigate a multiple working vacation machine repair system with $R(R \geq 2)$ homogeneous servers and mixed standbys operating under a multi-level threshold policy in this work. The remainder of this chapter has been structured in this manner: Section 1.2 discusses the generalized triadic policy model for the machine maintenance problem, as well as various notations and assumptions. In Section 1.3, the authors build the system governing steady-state equations using the birth-death rates and solve them recursively to obtain steady-state probabilities. A number of performance measures are established in Section 1.4. Section 1.5 illustrates the behavior of various performance metrics with different parameters and thresholds through numerical findings.

1.2 MODEL DESCRIPTION

A redundant machining system with M identical operating machines, S warm standbys, C cold standbys, and R servers is considered. The structure on a total has $L(= M + S + C)$ machines. If there are fewer than M operating machines in the system, it is said to be short. At initial stage, when no failed machine in the system, all R servers remain in dormant state. Each time the number of failed machines reaches $N_i(1 \leq i \leq R)$ (such that $N_i \leq N_j, \forall i < j; 1 \leq i < j \leq R$), one of the dormant $R - i + 1$ servers becomes active. Similarly, when the number of failed machines falls to $Q_i(1 \leq i \leq R - 1)$ (such that $Q_i > Q_j, \forall i < j; 1 \leq i < j \leq R - 1$), the server that was recently finishing a service among the concurrently active $R - i + 1$ servers becomes inactive. Furthermore, if the number of failed machines in the system goes to zero and meanwhile only one server is operational, the single active server gets removed. This is referred to as *generalized triadic policy*. Each time if there does not exist any failed machine in the structure, all the servers gets switched into the working vacation mode. At a working vacation completion epoch, if the count of failed machines accumulated in the repairable machining structure is N_1, then the ??-servers shift to regular busy mode and begin operating as per generalized triadic policy. In situation if this threshold is not met, the servers grab another working vacation and keep going with doing the same till the count of failed machines is N_1 or above toward the completion of a vacation period. This has been described as the *multiple working vacation policy*. The service times of R-servers in working vacation period and normal busy period are assumed as exponentially distributed with service rates μ and $\eta(\eta < \mu)$, respectively.

1.2.1 ASSUMPTIONS

The repairable redundant machining model under generalized triadic policy is mathematically formulated with the following assumptions:

1. The operating machines, warm standbys, and cold standbys are prone to failures in accordance with an exponential distribution having parameters λ, α, and *zero*, respectively.
2. When a machine in operation fails, it gets swapped by a warm standby if one is available. This warm standby is further substituted with an available cold standby. Warm standby acquires the failure characteristics of an operating machine when it transitions to an operating state, and a cold standby acquires the failure characteristics of a warm standby when it transitions to a warm standby state. The changeover in any of the above situation is instantaneous.
3. When the machine in operation or standby mode undergoes failure, it is instantly moved to a service facility and repaired in the chronological order of their failure. The servers in the service facility operate in two periods: one is the regular busy period, when at least one of the R-servers has returned from vacation, and each of the R-homogeneous servers provides repair (service) to the failed machines under generalized triadic policy with exponentially distributed service rate μ, and the other is the working vacation period, when all the servers are on vacation mode and only one of the R-servers will offer a service with exponentially distributed service rate $\eta(\eta < \mu)$.
4. Each server at a time can provide service to one machine only. If on arriving at the repair facility, a failed machine does not find any free server to provide service, it must join the queue in the system until a server is available.
5. After undergoing servicing, a failed machine is treated as fresh one and shifts to cold standby mode, provided there are less than M machines in operation mode, in which case the repaired machine is set into operation mode.

6. When each server discovers that no failed machines are waiting to be repaired, it goes on vacation mode.
7. The working vacation periods of the servers are independently and identically distributed at an exponential rate ω.
8. The state-dependent failure rate λ_n is defined as:

$$
\lambda_n = \begin{cases}
M\lambda + (S-n)\alpha & 1 \le n < C \\
M\lambda + (C+S-n)\alpha & C \le n < C+S \\
(L-n)\lambda & C+S \le n < L \\
0 & \text{otherwise}
\end{cases}
$$

1.2.2 NOTATIONS

The model has been formulated utilizing the subsequent notations:

M: Total count of operating machines in the system.
S: Total count of warm standby machines in the system.
C: Total count of cold standby machines in the system.
R: Total count of servers in the system.
λ: Failure rate of operating machines.
α: Failure rate of warm standby machines.
μ: Service rate of server during regular busy period.
η: Service rate of server during working vacation period.
ω: Working vacation rate of the server.

1.3 QUEUE SIZE DISTRIBUTION

The pair $\{(i,n) : 0 \le i \le R$ and $0 \le n \le L$ denotes the system state. Here $i = 0$ denotes the working vacation period and, $i = 2, 3, \ldots, R$ denotes the state when i $(1 \le i \le R)$ number of servers are active during the usual busy period. The number of failed machines in the system is represented by the index n.

The system that governs steady-state probabilities for various states of a repairable machining system is defined as follows:

$P_{0,n} \equiv$ The probability that the system is in working vacation mode with n failed machines, where $n = 0, 1, 2, \ldots, L$
$P_{i,n} \equiv$ The probability that the system is in the regular busy period mode with i active servers and n failed machines, where $i = 1, 2, 3, \ldots, R-1$. Here for $i = 1$, $n = 1, 2, 3, \ldots, N_2 - 1$ and for $i = 2, 3, \ldots, R-1$, $n = Q_{R-i+1} + 1, Q_{R-i+1} + 2, \ldots, Ni + 1 - 2, N_{i+1} - 1$
$P_{R,n} \equiv$ The probability that the system is in the regular busy period mode with R active servers and n failed machines, where $n = Q_1 + 1, Q_1 + 2, \ldots, L-1, L$

The following steady-state equations have been established using the aforementioned assumptions and notations for the model under study.

(i) When all the R-servers are on working vacation

$$\lambda_o P_{0,0} = \eta P_{0,1} + \mu P_{1,1} \,, \tag{1.1}$$

$$[\lambda_n + \eta]\, P_{0,n} = \eta P_{0,n+1} + \lambda_{n-1}P_{0,n-1}\,, \quad 1 \le n \le N_1 - 1\,, \tag{1.2}$$

$$[\lambda_n + \eta + \omega]\, P_{0,n} = \eta P_{0,n+1} + \lambda_{n-1}P_{0,n-1}\,, \quad N_1 \le n \le L - 1\,, \tag{1.3}$$

$$[\eta + \omega]\, P_{0,L} = \lambda_{L-1}P_{0,L-1}\,, \tag{1.4}$$

(ii) When one server is active during the regular busy mode

$$[\mu + \lambda_1]\, P_{1,1} = \mu P_{1,2}\,, \tag{1.5}$$

$$[\lambda_n + \mu]\, P_{1,n} = \mu P_{1,n+1} + \lambda_{n-1}P_{1,n-1}\,, \quad 2 \le n \le Q_{R-1} - 1\,, \tag{1.6}$$

$$\left[\lambda_{Q_{R-1}} + \mu\right]\, P_{1,Q_{R-1}} = \mu P_{1,Q_{R-1}+1} + \lambda_{Q_{R-1}-1}P_{1,Q_{R-1}-1} + 2\mu P_{2,Q_{R-1}+1}\,, \tag{1.7}$$

$$[\lambda_n + \mu]\, P_{1,n} = \mu P_{1,n+1} + \lambda_{n-1}P_{1,n-1}\,, \quad Q_{R-1} + 1 \le n \le N_1 - 1\,, \tag{1.8}$$

$$[\lambda_n + \mu]\, P_{1,n} = \mu P_{1,n+1} + \lambda_{n-1}P_{1,n-1} + \omega P_{0,n}\,, \quad N_1 \le n \le N_2 - 2\,, \tag{1.9}$$

$$\left[\lambda_{N_2-1} + \mu\right]\, P_{1,N_2-1} = \lambda_{N_2-2}P_{1,N_2-2} + \omega P_{0,N_2-1}\,, \tag{1.10}$$

(iii) When $i(2 \le i \le R - 1)$ servers are active during the regular busy mode

$$\left[\lambda_{Q_{R-i}+1} + i\mu\right]\, P_{i,Q_{R-2}+1} = i\mu P_{i,Q_{R-i}+2}\,, \tag{1.11}$$

$$[\lambda_n + i\mu]\, P_{i,n} = i\mu P_{i,n+1} + \lambda_{n-1}P_{i,n-1}\,, \quad Q_{R-i+1} + 2 \le n \le Q_{R-i} - 1\,, \tag{1.12}$$

$$\left[\lambda_{Q_{R-i}} + i\mu\right]\, P_{i,Q_{R-i}} = i\mu P_{i,Q_{R-i}+1} + \lambda_{Q_{R-i}-1}P_{i,Q_{R-i}-1} + (i+1)\mu P_{i+1,Q_{R-i}+1}\,, \tag{1.13}$$

$$[\lambda_n + i\mu]\, P_{i,n} = i\mu P_{i,n+1} + \lambda_{n-1}P_{i,n-1}\,, \quad Q_{R-i} + 1 \le n \le N_i - 1\,, \tag{1.14}$$

$$[\lambda_{N_i} + i\mu]\, P_{i,N_i} = i\mu P_{i,N_i+1} + \lambda_{N_i-1}P_{i,N_i-1} + \lambda_{N_i-1}P_{i-1,N_i-1} + \omega P_{0,N_i}\,, \tag{1.15}$$

$$[\lambda_n + i\mu]\, P_{i,n} = i\mu P_{i,n+1} + \lambda_{n-1}P_{i,n-1} + \omega P_{0,n}\,, \quad N_i + 1 \le n \le N_{i+1} - 2\,, \tag{1.16}$$

$$\left[\lambda_{N_{i+1}-1} + i\mu\right]\, P_{i,N_{i+1}-1} = \lambda_{N_{i+1}-2}P_{i,N_{i+1}-2} + \omega P_{0,N_{i+1}-1}\,, \tag{1.17}$$

(iv) When R servers are active during the regular busy mode

$$[\lambda_{Q_1+1} + R\mu]\, P_{R,Q_1+1} = R\mu P_{R,Q_1+2}\,, \tag{1.18}$$

$$[\lambda_n + R\mu]\, P_{R,n} = R\mu P_{R,n+1} + \lambda_{n-1}P_{R,n-1}\,, \quad Q_1 + 2 \le n \le N_R - 1\,, \tag{1.19}$$

$$[\lambda_{N_R} + R\mu]\, P_{R,N_R} = R\mu P_{R,N_R+1} + \lambda_{N_R-1}P_{R,N_R-1} + \lambda_{N_R-1}P_{R-1,N_R-1} + \omega P_{0,N_R}\,, \tag{1.20}$$

$$[\lambda_n + R\mu]\, P_{R,n} = R\mu P_{R,n+1} + \lambda_{n-1}P_{R,n-1} + \omega P_{0,n}\,, \quad N_R + 1 \le n \le L - 1\,, \tag{1.21}$$

$$R\mu P_{R,L} = \lambda_{L-1}P_{R,L-1} + \omega P_{0,L}\,, \tag{1.22}$$

The expressions for steady-state probabilities in terms of $P_{0,L}$ are computed employing the recursive method for the Equations (1.1)–(1.22). Then, we make use of the normalizing condition given below to obtain $P_{0,L}$.

$$\sum_{n=0}^{L} P_{0,n} + \sum_{n=1}^{N_2-1} P_{1,n} + \sum_{i=2}^{R-1}\sum_{n=Q_{R-i+1}+1}^{N_{i+1}-1} P_{i,n} + \sum_{n=Q_1+1}^{L} P_{R,n} = 1$$

Further, we utilize this assessed value of $P_{0,L}$ to attain final expressions for steady-state probabilities. In the following section, we investigate system behavior by developing formulations for certain significant performance measures based on the probabilities of various system states.

1.4 PERFORMANCE MEASURES

A critical component of the quantitative study of machining system modeling is the evaluation of system performance measures. In this section, we use steady-state probabilities to derive expressions for different performance measures including the average number of failed machines, the average number of busy and idle servers, throughput, and so on:

- $E(N_S) \equiv$ The average number of failed machines in the system is:

$$E(N_S) = \sum_{n=0}^{L} nP_{0,n} + \sum_{n=1}^{N_2-1} nP_{1,n} + \sum_{i=2}^{R-1} \sum_{n=Q_{R-i+1}+1}^{N_{i+1}-1} nP_{i,n} + \sum_{n=Q_1+1}^{L} nP_{R,n}$$

- $E(B_0) \equiv$ The expectation of one busy server being busy throughout the working vacation period is:

$$E(B_0) = \sum_{n=1}^{L} P_{0,n}$$

- $E(B_i) \equiv$ The expectation of $i(1 \leq i \leq R)$ servers being busy throughout the normal busy period is:

$$E(B_1) = \sum_{n=1}^{N_2-1} P_{1,n}$$

$$E(B_i) = \sum_{Q_{R-i+1}+1}^{N_{i+1}-1} iP_{i,n} \; ; \; 2 \leq i \leq R-1$$

$$E(B_R) = \sum_{Q_1+1}^{L} RP_{R,n}$$

- $E(I) \equiv$ The average number of idle servers in the system is:

$$E(I) = RP_{0,0} + \sum_{n=1}^{L} (R-1)P_{O,n} + \sum_{n=1}^{N_2-1} (R-1)P_{1,n} + \sum_{i=2}^{R-1} \sum_{n=Q_{R-i+1}+1}^{N_{i+1}-1} (R-i)P_{i,n}$$

- MA$\equiv$ Machine availability(the proportion of time that the machines are working):

$$\text{M.A.} = 1 - \frac{E(N_S)}{L}$$

- OU$\equiv$ Operative utilization(the proportion of busy servers):

$$\text{O.U.} = \frac{\sum_{i=0}^{R} E(B_i)}{R}$$

- Th. $\equiv$ Throughput(the average number of failed machines served):

$$\text{Th.} = \sum_{n=1}^{L} \eta P_{0,n} + \sum_{n=2}^{N_2-1} \mu P_{1,n} + \sum_{i=3}^{R-1} \sum_{n=Q_{R-i+1}+2}^{N_{i+1}-1} i\mu P_{i,n} + \sum_{n=Q_{R-1}+2}^{L} R\mu P_{R,n}$$

In the following section, some pertinent numerical findings for the various system parameters are obtained.

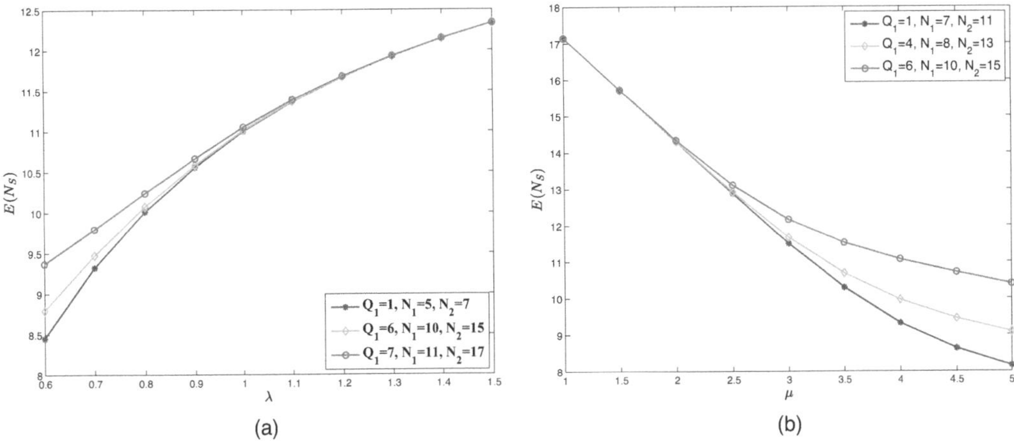

Figure 1.1 Variation of $E(N_S)$ with (a) λ and (b) μ. Here the other parameter values are assigned as $R = 2, L = 20; M = 17; C = 1; S = 2; \alpha = 0.3; \eta = 1.8; \omega = 1.0$.

1.5 NUMERICAL RESULTS

The analytical findings were accomplished by using a recursive method. Now, certain numerical computations in the form of graphs and tables are required to validate these results with a real-time queueing system. This is where the MATLAB® software of version **8.1.0.604 (R2013a)** comes in handy. The graphs showing the variation of several performance measures with different parameters are presented.

1.5.1 VARIATION OF AVERAGE NUMBER OF FAILED MACHINES IN THE SYSTEM $E(N_S)$

- Given that the failure rate λ has been increasing at a rapid pace, machines that fail arrive at the service facility in a relatively short period of time. Consequently, there is a significant increase in the average number of failing machines in the system, $E(N_S)$ as displayed in Figure 1.1a.
- An increase in the service rate μ results in a decrease in the repair time for failed machines. As a result, the average number of failed machines in the system, denoted as $E(N_S)$, begins to decrease. Visual verification for the aforementioned assertions is provided by Figure 1.1b.

1.5.2 VARIATION OF THE THROUGHPUT (TH)

Throughput is the average number of failed machines processed in a unit of time.

- Figures 1.2a and 1.3a illustrate the trend of throughput across various values of λ. The throughput demonstrates an increasing trend with the increase in λ until it reaches a certain threshold, beyond which it starts to decline, as depicted in Figure 1.3a. This behavior is observed due to the increase in failure rate, which leads to the quick attainment of activation thresholds. This action enhances the probability of additional servers being in an active mode. Therefore, the average number of failed machines processed within a given time period will increase, resulting in an increase in the throughput. As the failure rate of machines increases further, there is a higher probability of the repair facility experiencing increased workload and potential congestion. In the event that the repair facility experiences high levels of activity, it is possible that the throughput could be constrained.
- With increase in the number of active servers, the number of failed machines processed per unit time will also increase and thereby the system throughput will increase as shown in Figure

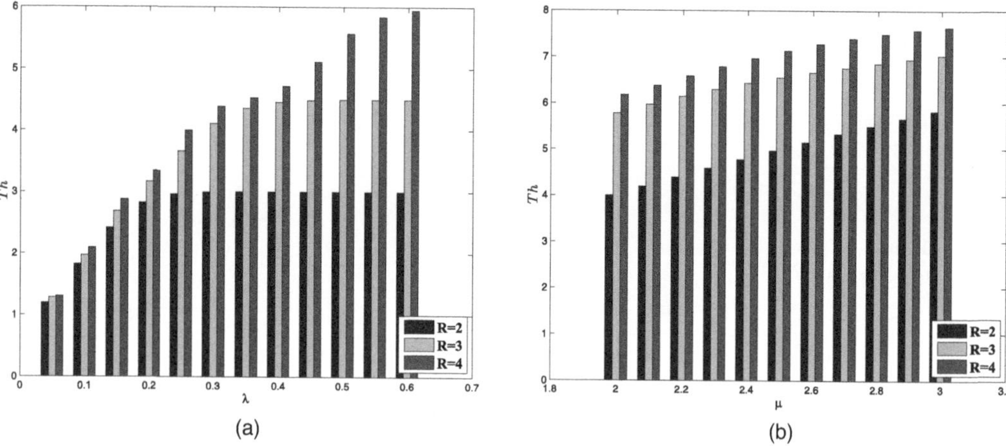

Figure 1.2 Variation of throughput (Th) with λ (a) and μ (b) for different R values.

1.2. When the service rate μ increases, the average number of failed machines repaired per unit of time also increases. The increase in machine repairs and reduced busyness of the repair facility contribute to an increase in system throughput as demonstrated Figures 1.2b and 1.3b.

- Figure 1.3b indicates that with the increase in activation thresholds (N_1, N_2), the likelihood of more servers being in active mode decreases. Thus, the service process will be slow leading to congestion and hence the system throughput decreases.
- As the service rate of the server during working vacation(η) increases, the tendency of system being in working vacation mode increases and hence the service facility continues to be working with reduced service rate. This leads to decrease in system throughput as displayed in Figure 1.3c.

1.5.3 VARIATION IN THE EXPECTATION OF BUSY SERVERS AND THE AVERAGE NUMBER OF IDLE SERVERS

The change in the nature of the proportion of average busy servers $(E(B_0), E(B_1), E(B_2))$ and idle servers $(E(I))$ with service rate is presented in Figure 1.4.

- Given that the count of failed machines in the structure begins to diminish the service rate μ(or η) increases, the possibility of both servers being in an active state reduces. As a result, in this scenario, the value of $E(B_2)$ will decrease, whereas the values of $E(I), E(B_0)$, and $E(B_1)$ will increase, as can be seen in Figure 1.3.
- Increasing the thresholds N_i reduces the likelihood of all the servers being in active mode. Therefore, $E(B_2)$ decreases while $E(B_0)$ and $E(B_1)$ increase. Raising the deactivation threshold values will decrease the likelihood of the service facility being in working vacation mode, thereby decreasing $E(B_0)$. Due to short chances of less servers being in active mode, the average number of idle servers $E(I)$ increases.

1.5.4 VARIATION IN THE MACHINE AVAILABILITY (MA) AND THE OPERATIVE UTILIZATION (OU)

Table 1.1 shows the effect of parameters η, μ, and α on performance measures such as MA and OU. Table 1.2 displays the change in the behavior of various performance measures such as MA and OU with change in parameters λ, μ thresholds (Q_1, N_1, N_2) and (L, M, S, C) values.

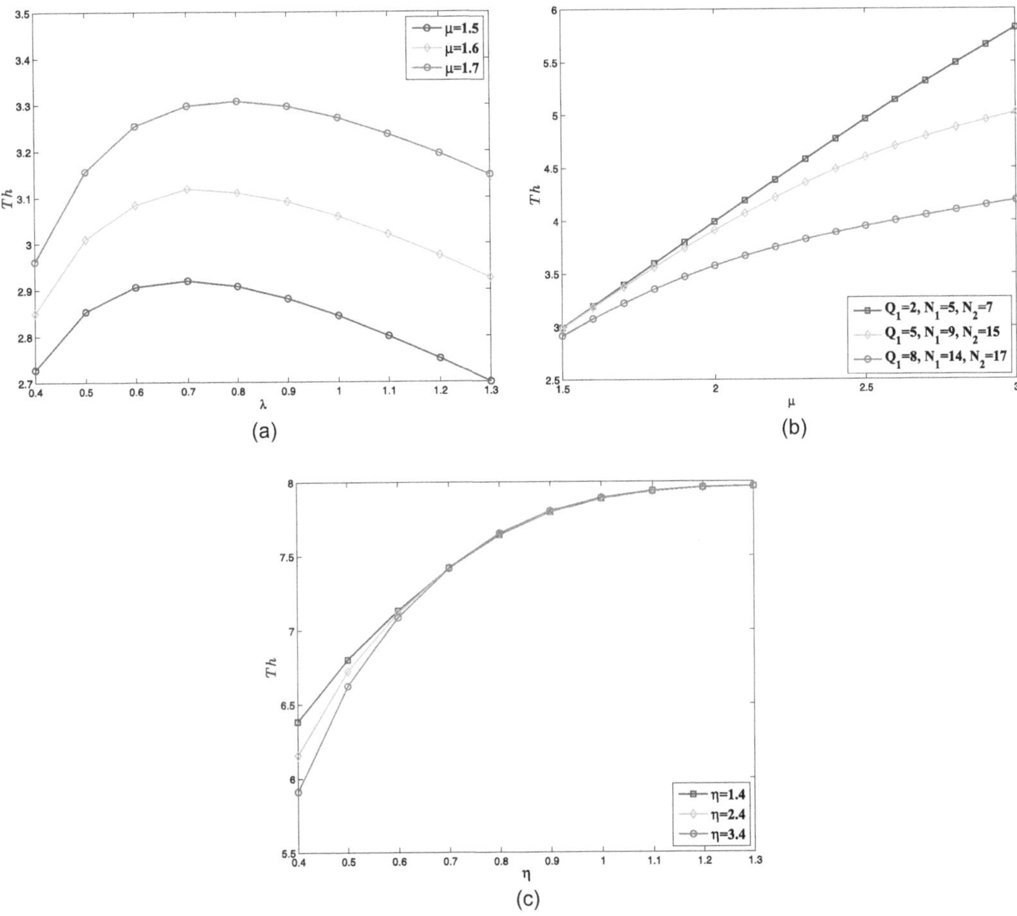

Figure 1.3 Variation of throughput with (a) λ, (b) μ and (c) η. Here the other parameter values are assigned as $R = 2, L = 20; M = 17; C = 1; S = 2; \alpha = 0.3; \eta = 1.8; \omega = 1.0$.

Table 1.1

Performance Measures with Parameters α, η, and ω for $R=2$

η	α	$\omega = 0.05$		$\omega = 0.07$		$\omega = 0.09$	
		MA	OU	MA	OU	MA	OU
	0.02	0.5324	0.7930	0.5366	0.8006	0.5390	0.8048
1.0	0.05	0.5329	0.7945	0.5370	0.8018	0.5393	0.8075
	0.08	0.5335	0.7960	0.5373	0.8029	0.5395	0.8066
	0.02	0.5330	0.7927	0.5371	0.8003	0.5394	0.8045
1.1	0.05	0.5336	0.7942	0.5374	0.8015	0.5397	0.8054
	0.08	0.5341	0.7957	0.5378	0.8026	0.5399	0.8064
	0.02	0.5336	0.7923	0.5376	0.8000	0.5398	0.8041
1.2	0.05	0.5342	0.7939	0.5379	0.8012	0.5400	0.8051
	0.08	0.5346	0.7954	0.5382	0.8023	0.5402	0.8061

Table 1.2
Performance Measures with (*L,M,S,C*) Values Parameters λ , μ and Thresholds ($Q(1),N(1),N(2)$) for $R=2$

L,M,S,C	$Q(1),N(1),N(2)$	λ	$\mu = 2.0$		$\mu = 2.1$		$\mu = 2.2$	
			M.A.	O.U.	M.A.	O.U.	M.A.	O.U.
		0.14	0.8196	0.8285	0.8236	0.8060	0.8319	0.7481
	1,3,5	0.17	0.7797	0.9094	0.7905	0.8925	0.7993	0.8753
		0.20	0.7279	0.9557	0.7462	0.9442	0.7599	0.9318
		0.14	0.7001	0.5677	0.9094	0.5411	0.7188	0.5224
25,18,5,2	3,5,14	0.17	0.6664	0.6709	0.6744	0.6437	0.6821	0.6185
		0.20	0.6379	0.7664	0.6469	0.7373	0.6550	0.7098
		0.14	0.6049	0.5079	0.6199	0.4918	0.6347	0.4765
	5,10,18	0.17	0.5586	0.5808	0.5679	0.5606	0.5781	0.5427
		0.20	0.5362	0.6605	0.5418	0.6346	0.5480	0.6115
		0.14	0.8015	0.9197	0.8110	0.9042	0.8188	0.8884
	1,3,5	0.17	0.7469	0.9672	0.7627	0.9579	0.7759	0.9478
		0.20	0.6803	0.9883	0.7021	0.9836	0.7213	0.9780
		0.14	0.7005	0.6740	0.7077	0.6470	0.7145	0.6221
28,23,3,2	3,5,14	0.17	0.6689	0.7877	0.6776	0.7587	0.6852	0.7312
		0.20	0.6320	0.8798	0.6446	0.8535	0.6553	0.8272
		0.14	0.6091	0.5923	0.6175	0.5710	0.6267	0.5526
	5,10,18	0.17	0.5836	0.6922	0.5892	0.6651	0.5984	0.6404
		0.20	0.5647	0.7893	0.5708	0.7595	0.5762	0.7315
		0.14	0.8064	0.8484	0.8126	0.8282	0.8179	0.8082
	1,3,5	0.17	0.7721	0.9186	0.7812	0.9040	0.7889	0.8890
		0.20	0.7317	0.9579	0.7449	0.9481	0.7560	0.9375
		0.14	0.6800	0.5384	0.6908	0.5193	0.7018	0.5020
23,20,2,1	3,5,14	0.17	0.6443	0.6247	0.6521	0.6013	0.6600	0.5801
		0.20	0.6207	0.7106	0.6276	0.6837	0.6341	0.6589
		0.14	0.6080	0.4814	0.6252	0.4692	0.6409	0.4567
	5,10,18	0.17	0.5427	0.5281	0.5574	0.5158	0.5726	0.5048
		0.20	0.5059	0.5809	0.5149	0.5628	0.5251	0.5474

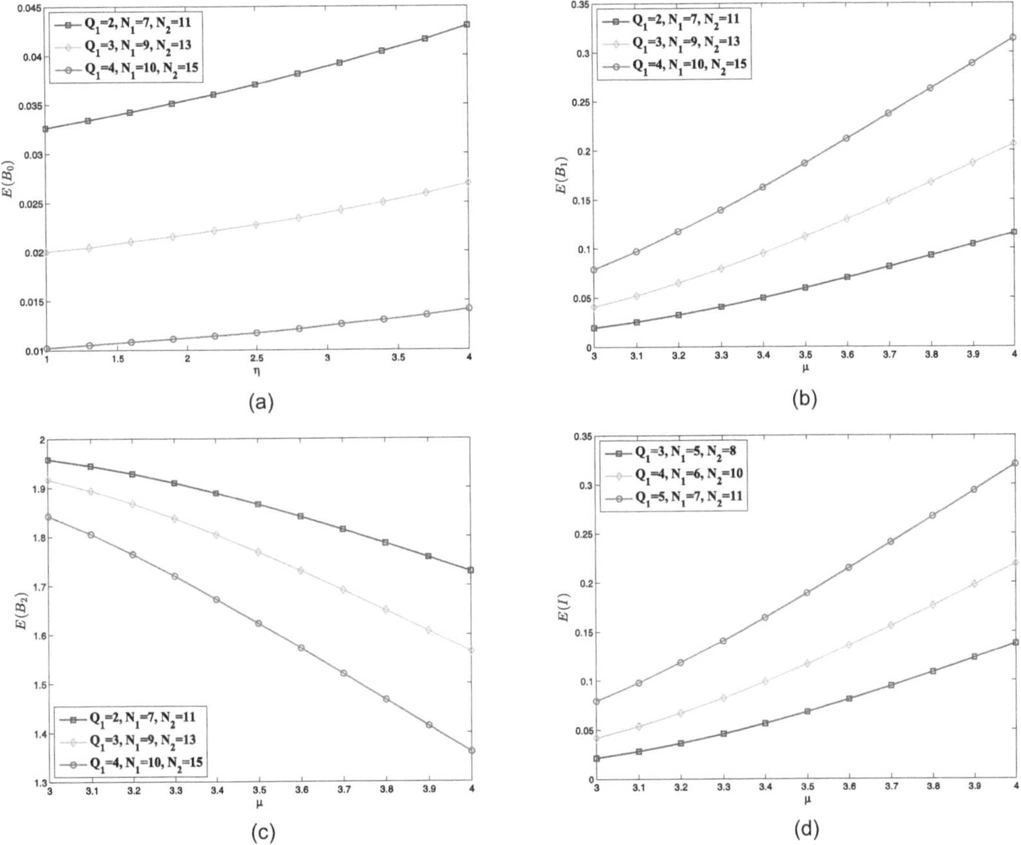

Figure 1.4 Variation of $(E(B_0), E(B_1), E(B_2))$ and $(E(I))$ with, μ and η. Here the other parameter values are assigned as $R = 2, L = 24; M = 18; C = 2; S = 4; \alpha = 0.3; \lambda = 0.5; \omega = 1.0$. (a) Variation of $(E(B_0))$ with η, (b) variation of $(E(B_1))$ with μ, (c) variation of $(E(B_2))$ with μ, and (d) variation of $(E(I))$ with μ.

- With a rise in failure rates (α, λ), the chances of attainment of server activation thresholds increase further leading to an increase in the expectation of more servers being in active state during the regular busy period. Thus, operative utilization increases. Higher failure rates imply more frequent breakdowns of machines in the system, resulting in longer downtime and less machine availability.

- An increase in service rates (μ, η) leads to faster repairs of failed machines, hence shorter downtime and higher availability of machines. With a rise in the service rates, the possibility of all servers being in an active state reduces but $E(B_i); 1 \leq i \leq R - 1$ will increase. Thus, operative utilization will also increase.

- With a rise in the working vacation rate (ω), the time duration of working vacation decreases. Thus, the failed machines will be repaired quickly leading to a rise in machine availability in the system. This will also enhance the chances of system being in regular busy period and more servers being in active mode, thus improving operative utilization.

- The rise in activation thresholds decreases the chances of more servers being in active mode, which slows the service process and further leads to accumulation of failed machines. Therefore, both machine availability and service utilization decrease.

1.6 CONCLUSION

Our study is distinctive in that it seeks to generalize triadic control policy for an R-server redundant machining system with multiple server vacations. Using a recursive technique, we arrived at an analytic conclusion for the system regulating probabilities. Numerical computations in the form of graphs and tables are demonstrated for analyzing system performance measures in relation to some system parameters. This problem can be expanded by including some more realistic characteristics in the queueing modeling of a generalized triadic policy for a machine repair problem, such as impatient behavior of failed machines and server unreliability.

ACKNOWLEDGMENT

Without the remarkable support of my supervisor, Dr. Kamlesh Kumar, this investigation simply could not have been carried out. In addition, I would like to express my appreciation to my fellow research scholar, Shalini Sharma, for the contribution she has made during the entirety of this project.

REFERENCES

1. Bell, C. E.: Optimal operation of an $M/M/2$ queue with removable servers. *Oper. Res.* **28**(5): 1189–1204 (1980). https://doi.org/10.1287/opre.28.5.1189.
2. Gao, S. and Wang, J.: Reliability and availability analysis of a retrial system with mixed standbys and an unreliable repair facility. *Reliab. Eng. Syst. Saf.* **205**: 107240 (2021). https://doi.org/10.1016/j.ress.2020.107240.
3. Gnedenko, B. V., Belyayev, Y. K., and Solovyev, A. D.: *Mathematical Methods of Reliability Theory.* New York: Academic Press (1969).
4. Haung, H. I., Hsu, P. C., and Ke, J. C.: Controlling arrival and service of a two-removable-server system using genetic algorithm. *Expert Syst. Appl.* **38**: 10054–10059 (2011). https://doi.org/10.1016/j.eswa.2011.02.011.
5. Ke, J. C., Wu, C. H., and Pearn, W. L.: Dynamic operating policy for the controllable queue with two removable unreliable servers. *Int. J. Comput. Math.* **2**: 81–96 (2017). https://doi.org/10.1080/23799927.2017.1330281.
6. Ketema, T., Demie, S., and Tefera, M.: Controllable $M/M/2$ machine repair problem with multiple working vacations and triadic $(0,Q,N,M)$ policy. *Int. J. Manag. Sci. Eng. Manag.* **16**: 1–13 (2021). https://doi.org/10.1080/17509653.2021.1935358.
7. Kumar, K. and Jain, M.: Threshold $N-$policy for (M,m) degraded machining system with $K-$heterogeneous servers, standby switching failure and multiple vacations. *Int. J. Math. Oper.* **5**(4): 423–425 (2013).
8. Kumar, K., Jain, M., and Shekhar, C.: Machine repair system with $F-$policy, two unreliable servers, and warm standbys. *J. Test. Eval.* **47**(1):361–383 (2019). https://doi.org/10.1520/JTE20160595. ISSN 0090-3973
9. Levy, Y. and Yechiali, U.: Utilization of the idle time in an $M/G/1$ queue. *Manag. Sci.* **22**: 202–211 (1975). https://doi.org/10.1287/mnsc.22.2.202.
10. Lin, C. H. and Ke, J. C.: Genetic algorithm for optimal thresholds of an infinite capacity multi-server system with triadic policy. *Expert Syst. Appl.* **37**: 4276–4282 (2010). https://doi.org/10.1016/j.eswa.2009.11.074.
11. Rhee, H. K. and Sivazlian, B. D.: Distribution of the busy period in a controllable $M/M/2$ queue operating under the triadic $(0,K,N,M)$ policy. *J. Appl. Prob.* **27**: 425–432 (1990). https://doi.org/10.2307/3214662.
12. Servi, L. D. and Finn, S. G.: $M/M/1$ queue with working vacations $(M/M/1/WV)$. *Perform. Eval.* **50**: 41–52(2002). https://doi.org/10.1016/S0166-5316(02)00057-3.
13. Taylor, J. and Jackson, H. R. P.: Application of birth and death processes to provision of spare machines. *J. Oper. Res. Soc.* **5**: 95–108 (1954). https://doi.org/10.2307/3007087.

14. Wang, K. H. and Chang, C. L.: Reliability of a controllable $M/M/2$ system with warm standbys operating under the triadic $(0, Q, N, M)$ policy. *Compu. Ind. Eng.* **28**: 163–178 (1995). https://doi.org/10.1016/0360-8352(94)00035-L.

15. Wang, K. H. and Wang, Y. L.: Optimal control of an $M/M/2$ queueing system with finite capacity operating under the triadic $(0, Q, N, M)$ policy. *Math. Meth. Oper. Res.* **55**: 447–460 (2002). https://doi.org/10.1007/s001860200190.

16. Wu, D. and Takagi, H.: $M/G/1$ queue with multiple working vacations. *Perform. Eval.* **63**: 654–681 (2006). https://doi.org/10.1016/j.peva.2005.05.005.

17. Yadin, M. and Naor, P.: Queueing systems with a removable service station. *J. Oper. Res. Soc.* **14**: 393–405 (1963). https://doi.org/10.1057/jors.1963.63.

2 Solving Physical System Problems Using Partial Differential Equations

Sahil Bansal

2.1 INTRODUCTION

A partial differential equation includes derivatives of an unknown function with respect to two or more independent variables. To solve a physical system and to study its behavior, formulation of the system in terms of mathematical equations involving the system variables is required. The equations are then solved by applying various mathematical tools depending on the nature of system and the boundary/working conditions [1,3]. Several physical problems can be formulated in terms of partial differential equations. This chapter reviews and describes some of the partial differential equations that are often used in physics to study the physical systems [1]. Given below are some of the partial differential equations along with the nature of physical problems in which they are used.

2.1.1 POISSON'S EQUATION

The differential equation is

$$\nabla^2 \phi = \rho$$

Here, ρ is called source density, e.g., charge density in case of electric field and mass density in case of gravitational field, and is dependent on position co-ordinates (x, y, z). ϕ is a scalar potential function, e.g., electric potential in case of electric field and gravitational potential in case of gravitational field. This equation can be used to find electric potential in a region of charge, gravitational potential in a region containing mass distribution, magnetic potential in case of magnetic distribution, temperature distribution in case of heat flow, etc.

2.1.2 LAPLACE EQUATION

This is a special case of Poisson Equation. The equation is

$$\nabla^2 \phi = 0$$

Here, ϕ is a scalar potential, which is the same as the one in Poisson equation. The equation can be used to find electric potential inside a uniform dielectric, gravitational potential in region of no mass, electric potential in charge free space, magnetic potential in current free space, etc.

2.1.3 WAVE EQUATION

$$\nabla^2 y = \frac{1}{c^2} \frac{\partial^2 y}{\partial t^2}$$

DOI: 10.1201/9781003407386-2

$$\implies \nabla^2 y - \frac{1}{c^2}\frac{\partial^2 y}{\partial t^2} = 0 \quad \text{or} \quad \Box^2 y = 0$$

where $\Box^2 = \nabla^2 - \frac{1}{c^2}\frac{\partial^2 y}{\partial t^2}$ is called D'Alembertian. y can be the displacement of a stretched string, the displacement of a particle of medium as wave propagates, the current in electrical transmission line, etc., depending on the nature of physical problem. c is velocity of wave which is related to the density of the medium and its elasticity.

2.1.4 SCHRODINGER'S WAVE EQUATION

This equation is the fundamental equation of quantum mechanics. The equation in time-dependent form and time independent form, respectively, can be written as

$$-\frac{\hbar^2}{2m}\nabla^2\Psi = \iota\hbar\frac{\partial\Psi}{\partial t} - V\Psi$$

$$\nabla^2\Psi + \frac{2m}{\hbar^2}(E - V)\Psi = 0$$

E is the total energy of the particle, and V is the potential energy of the particle. Ψ is the wave function corresponding to the particle of mass m and $\hbar = \frac{h}{2\pi}$ with 'h' being the Planck's constant. The equation finds an immense use in solving quantum mechanical systems.

2.1.5 EQUATION OF HEAT FLOW

The equation for flow of heat in time-dependent form is

$$D^2(\nabla^2\phi) = \frac{\partial\phi}{\partial t}$$

Here, D^2 is called diffusivity and is constant for a given medium. ϕ is the concentration of diffusing material. The equation is used in modeling the heat flow through the medium.

2.2 SOLUTIONS OF PARTIAL DIFFERENTIAL EQUATIONS

When a given function depends on a number of independent variables, the partial differential equation can be solved by the separation of variables [1]. The variables may involve independent space and time co-ordinates. For example, if f depends on spatial coordinates x, y, and z, then we can substitute $f = XYZ$, such that X is a function of x only, Y is a function of y only, and Z is a function of z only, in the given differential equation and can subsequently form three simpler differential equations as illustrated below [1].

Illustration 2.1
Solve Helmholtz differential equation, i.e., $(\nabla^2 + k^2)u = 0$, by the method of separation of variables [2].

Solution: The equation in Cartesian coordinates can be written as

$$\left(\frac{\partial^2}{\partial x^2} + \frac{\partial^2}{\partial y^2} + \frac{\partial^2}{\partial z^2} + k^2\right)u = 0 \tag{2.1}$$

Here, u is a function of x, y, and z. So, we can substitute $u = XYZ$, where X, Y, and Z are functions of x only, y only, and z only, respectively.

Hence, Equation (2.1) can be written as

$$\frac{\partial^2 (XYZ)}{\partial x^2} + \frac{\partial^2 (XYZ)}{\partial y^2} + \frac{\partial^2 (XYZ)}{\partial z^2} + k^2 (XYZ) = 0$$

Divide this equation by XYZ throughout to obtain

$$\frac{1}{X}\frac{\partial^2 X}{\partial x^2} + \frac{1}{Y}\frac{\partial^2 Y}{\partial y^2} + \frac{1}{Z}\frac{\partial^2 Z}{\partial x^2} + k^2 = 0$$

$$\implies \frac{1}{Y}\frac{\partial^2 Y}{\partial y^2} + \frac{1}{Z}\frac{\partial^2 Z}{\partial z^2} + k^2 = -\frac{1}{X}\frac{\partial^2 X}{\partial x^2} \tag{2.2}$$

Now the left side of the equation is a function of y and z, while the right side of the equation is a function of x only; so, for the equation to be correct, each side must be equal to a constant, say k_1^2.

$$\therefore \quad -\frac{1}{X}\frac{\partial^2 X}{\partial x^2} = k_1^2 \tag{2.3}$$

$$\text{and} \quad \frac{1}{Y}\frac{\partial^2 Y}{\partial y^2} + \frac{1}{Z}\frac{\partial^2 Z}{\partial z^2} + k^2 = k_1^2 \tag{2.4}$$

Equation (2.4) can be re-written as

$$\frac{1}{Z}\frac{\partial^2 Z}{\partial z^2} + k^2 - k_1^2 = \frac{1}{Y}\frac{\partial^2 Y}{\partial y^2} \tag{2.5}$$

Now, LHS is a variable of z only, and RHS is a function of y only. For this equation to be correct, each side must be equal to a constant, say k_2^2

$$\therefore \quad -\frac{1}{Y}\frac{\partial^2 Y}{\partial y^2} = k_2^2 \tag{2.6}$$

$$\text{and} \quad \frac{1}{Z}\frac{\partial^2 Z}{\partial z^2} + k^2 - k_1^2 = k_2^2 \tag{2.7}$$

Now, Equations (2.3), (2.6), and (2.7) are three separate partial differential equations with each equation having only one variable; they can be written, respectively, as

$$\frac{1}{X}\frac{\partial^2 X}{\partial x^2} + k_1^2 X = 0 \tag{2.8}$$

$$\frac{1}{Y}\frac{\partial^2 Y}{\partial y^2} + k_2^2 Y = 0 \tag{2.9}$$

$$\text{and} \quad \frac{1}{Z}\frac{\partial^2 Z}{\partial z^2} + k_3^2 Z = 0 \tag{2.10}$$

where $k_3^2 = k^2 - k_1^2 - k_2^2$. Equations (2.8)–(2.10) are ordinary differential equations, and their solutions are

$$X = a_1 \cos k_1 x + b_1 \sin k_1 x, \, Y = a_2 \cos k_2 y + b_2 \sin k_2 y, \text{ and } Z = a_3 \cos k_3 z + b_3 \sin k_3 z$$

Illustration 2.2
Solve Laplace equation, i.e., $\nabla^2 \phi = 0$, by the method of separation of variables.

Solution: The equation in Cartesian coordinates can be written as

$$\left(\frac{\partial^2}{\partial x^2} + \frac{\partial^2}{\partial y^2} + \frac{\partial^2}{\partial z^2}\right)\phi = 0 \tag{2.11}$$

Here, ϕ is a function of x, y, and z. So, we can substitute $\phi = XYZ$, where X, Y, and Z are functions of x only, y only, and z only, respectively. Hence, Equation (2.11) can be written as

$$\frac{\partial^2(XYZ)}{\partial x^2} + \frac{\partial^2(XYZ)}{\partial y^2} + \frac{\partial^2(XYZ)}{\partial z^2} = 0 \tag{2.12}$$

Divide this equation by XYZ throughout to obtain

$$\frac{1}{X}\frac{\partial^2(X)}{\partial x^2} + \frac{1}{Y}\frac{\partial^2(Y)}{\partial y^2} + \frac{1}{Z}\frac{\partial^2(Z)}{\partial z^2} = 0$$

$$\implies \frac{1}{Y}\frac{\partial^2 Y}{\partial y^2} + \frac{1}{Z}\frac{\partial^2 Z}{\partial z^2} = -\frac{1}{X}\frac{\partial^2 X}{\partial x^2} \tag{2.13}$$

Now the left side of the equation is a function of y and z, while the right side of the equation is a function of x only; so, for the equation to be correct, each side must be equal to a constant, say k_1^2

$$\therefore \quad \frac{1}{X}\frac{\partial^2 X}{\partial x^2} = k_1^2 \tag{2.14}$$

$$and \quad -\frac{1}{Y}\frac{\partial^2 Y}{\partial y^2} - \frac{1}{Z}\frac{\partial^2 Z}{\partial z^2} = k_1^2 \tag{2.15}$$

Equation (2.15) can be re-written as

$$-\frac{1}{Z}\frac{\partial^2 Z}{\partial z^2} - k_1^2 = \frac{1}{Y}\frac{\partial^2 Y}{\partial y^2} \tag{2.16}$$

Now, LHS is a variable of z only and RHS is a function of y only. For this equation to be correct, each side must be equal to a constant, say k_2^2

$$\therefore \quad \frac{1}{Y}\frac{\partial^2 Y}{\partial y^2} = k_2^2 \tag{2.17}$$

$$and \quad -\frac{1}{Z}\frac{\partial^2 Z}{\partial z^2} - k_1^2 = k_2^2 \tag{2.18}$$

Now, Equations (2.14), (2.17), and (2.18) are three separate partial differential equations with each equation having only one variable, and they can be written, respectively, as

$$\frac{1}{X}\frac{\partial^2 X}{\partial x^2} - k_1^2 X = 0 \tag{2.19}$$

$$\frac{1}{Y}\frac{\partial^2 Y}{\partial y^2} - k_2^2 Y = 0 \tag{2.20}$$

$$and \quad \frac{1}{Z}\frac{\partial^2 Z}{\partial z^2} - k_3^2 Z = 0 \tag{2.21}$$

where $k_3^2 = -(k_1^2 + k_2^2)$. Equations (2.19)–(2.21) are ordinary differential equations, and their solutions are

$$X = a\, e^{k_1 x}, \quad Y = b\, e^{k_2 y} \text{ and } Z = c\, e^{k_3 z}$$

Hence, the general solution of Laplace equation becomes

$$\phi = abc\, e^{k_1 x} e^{k_2 y} e^{k_3 z}$$

$$\therefore \qquad \phi = abc\, e^{k_1 x + k_2 y + k_3 z}$$

Illustration 2.3

Solve the three-dimensional heat flow equation, i.e., $D^2(\nabla^2 \phi) = \frac{\partial \phi}{\partial t}$ by the method of separation of variables [3].

Solution: The heat flow equation is

$$D^2(\nabla^2 \phi) = \frac{\partial \phi}{\partial t} \tag{2.22}$$

Let the solution of Equation (2.22) is

$$\phi(x,y,z,t) = u(x,y,z)T(t) \tag{2.23}$$

Using Equation (2.23) in (2.22)

$$TD^2(\nabla^2 u) = u\frac{\partial T}{\partial t}$$

Divide by uT on both sides

$$\frac{1}{u}(\nabla^2 u) = \frac{1}{D^2 T}\frac{\partial T}{\partial t} \tag{2.24}$$

Left hand side of this equation is a function of space co-ordinates (x,y,z) only, and right hand side of this equation is the function of time only. For the equation to be correct, each side must be a constant, $-k^2$ (say). The constant is taken negative because the temperature of the body decreases with time.

$$\therefore \qquad \frac{1}{u}\nabla^2 u = -k^2$$

$$\implies \qquad \nabla^2 u + k^2 u = 0 \tag{2.25}$$

Also

$$\frac{1}{D^2 T}\frac{\partial T}{\partial t} = -k^2$$

$$\implies \qquad \frac{\partial T}{\partial t} = -k^2 D^2 T \tag{2.26}$$

Equation (2.25) is nothing but the Helmholtz equation which has been solved in previous illustration, by the method of separation of variables. Equation (2.26) can be rewritten as

$$\frac{\partial T}{T} = -k^2 D^2 \partial t \tag{2.27}$$

Integrating this equation, we get

$$\implies \qquad \log T = -k^2 D^2\, t + \log A \tag{2.28}$$

Here, $\log A$ is the constant of integration. Equation (2.28) can be solved (by integrating) to obtain T, as $T = A\, e^{-k^2 D^2 t}$. Hence, the complete solution of heat equation in three dimensions is

$$\phi(x,y,z,t) = u(x,y,z)\, A\, e^{-k^2 D^2 t}$$

2.3 SOLUTIONS OF PARTIAL DIFFERENTIAL EQUATIONS FOR SOME PHYSICAL SYSTEMS

Solving a physical system requires the exact knowledge of the boundary conditions and the corresponding variables which influence the system and its operation. Applying the boundary conditions, the system can be transformed into partial differential equations in terms of the respective variables of the system, which can then be solved to find the exact solution [4].

Given below are some examples of the physical system solutions in terms of partial differential equations [1].

2.3.1 VIBRATING STRING

A string is a uniform, flexible rod with large length with respect to its diameter. The equation of a transverse wave on a string along x-axis (wave equation) is given by [2]

$$\frac{\partial^2 y}{\partial x^2} = \frac{1}{c^2}\frac{\partial^2 y}{\partial t^2} \tag{2.29}$$

The displacement y depends on x and t, i.e., $y = y(x,t)$ Let us assume

$$u = x + ct \qquad \text{and} \qquad v = x - ct \tag{2.30}$$

then

$$y = y(u,v) \tag{2.31}$$

Clearly,

$$\frac{\partial u}{\partial x} = \frac{\partial u}{\partial x} = 1 \tag{2.32}$$

$$\frac{\partial u}{\partial t} = -\frac{\partial v}{\partial t} = c \tag{2.33}$$

Hence,

$$\frac{\partial y}{\partial x} = \frac{\partial y}{\partial u}\frac{\partial u}{\partial x} + \frac{\partial y}{\partial v}\frac{\partial v}{\partial x} = \frac{\partial y}{\partial u}\cdot 1 + \frac{\partial y}{\partial v}\cdot 1 = \left(\frac{\partial}{\partial u} + \frac{\partial}{\partial v}\right)y \tag{2.34}$$

$$\implies \quad \frac{\partial}{\partial x} = \frac{\partial}{\partial u} + \frac{\partial}{\partial v} \tag{2.35}$$

Similarly,

$$\frac{\partial y}{\partial t} = \frac{\partial y}{\partial u}\frac{\partial u}{\partial t} + \frac{\partial y}{\partial v}\frac{\partial v}{\partial t} = \frac{\partial y}{\partial u}\cdot c + \frac{\partial y}{\partial v}\cdot(-c) = c\left(\frac{\partial}{\partial u} - \frac{\partial}{\partial v}\right)y \tag{2.36}$$

$$\implies \quad \frac{\partial}{\partial t} = c\left(\frac{\partial}{\partial u} - \frac{\partial}{\partial v}\right) \tag{2.37}$$

Differentiate Equation (2.34) with respect to x, partially, to obtain

$$\frac{\partial^2 y}{\partial x^2} = \frac{\partial}{\partial x}\left(\frac{\partial y}{\partial x}\right) = \left(\frac{\partial}{\partial u} + \frac{\partial}{\partial v}\right)\left(\frac{\partial y}{\partial u} + \frac{\partial y}{\partial v}\right) = \frac{\partial^2 y}{\partial u^2} + \frac{\partial^2 y}{\partial v^2} + 2\frac{\partial^2 y}{\partial u \partial v} \tag{2.38}$$

Similarly, differentiate Equation (2.36) with respect to t, partially, to obtain

$$\frac{\partial^2 y}{\partial t^2} = \frac{\partial}{\partial t}\left(\frac{\partial y}{\partial t}\right) = c\left(\frac{\partial}{\partial u} - \frac{\partial}{\partial v}\right)c\left(\frac{\partial y}{\partial u} - \frac{\partial y}{\partial v}\right) = c^2\frac{\partial^2 y}{\partial u^2} + \frac{\partial^2 y}{\partial v^2} - 2\frac{\partial^2 y}{\partial u \partial v} \tag{2.39}$$

Using Equations (2.38) and (2.39), Equation (2.29) implies

$$\frac{\partial^2 y}{\partial u^2} + \frac{\partial^2 y}{\partial v^2} + 2\frac{\partial^2 y}{\partial u \partial v} = \frac{1}{c^2}c^2\left(\frac{\partial^2 y}{\partial u^2} + \frac{\partial^2 y}{\partial v^2} - 2\frac{\partial^2 y}{\partial u \partial v}\right)$$

$$\implies \quad 4\frac{\partial^2 y}{\partial u \partial v} = 0 \quad \text{or} \quad \frac{\partial}{\partial u}\left(\frac{\partial y}{\partial v}\right) = 0 \tag{2.40}$$

Integrating equation (2.40) with respect to u, we get

$$\frac{\partial y}{\partial v} = f(v), \quad \text{where} f(v) \text{ is the constant of integration}$$

Integrating this again with respect to v, we get

$$y = F_1(u) + \int f(v)dv = F_1(u) + F_2(v)$$

where F_1 and F_2 are arbitrary functions of u and v, respectively. Substituting values of u and v, we get

$$y = F_1(x + ct) + F_2(x - ct)$$

This is the D' Alembert's solution of vibrating string.

2.3.2 PROPAGATION OF SOUND IN A GASEOUS MEDIUM

A gaseous medium is a fluid, and so, we apply the equation of continuity of a fluid of density ρ and velocity v [1].

$$\vec{\nabla} \cdot (\rho \vec{v}) = -\frac{\partial \rho}{\partial t} \tag{2.41}$$

Using Euler's equation of motion

$$(\vec{v} \cdot \vec{\nabla})\vec{v} + \frac{\partial v}{\partial t} = \vec{F} - \frac{1}{\rho}\vec{\nabla}p \tag{2.42}$$

Here, p is the pressure of fluid and $\vec{F}$ is the body force per unit mass of fluid. In a gaseous medium, we can assume [1] that (i) the velocities are small, and so, only lower powers are to be considered;(ii) the body force is neglected as there are no inter-molecular forces; (iii) the motion is irrotational, and hence, a velocity potential ϕ exists such that

$$\vec{v} = -\vec{\nabla}\phi \tag{2.43}$$

Defining, $\rho = \rho_0(1 + k)$, where ρ_0 is the initial density and k is the condensation term. k is extremely small for a gas , because as sound propagates there is a negligible change in the density of gas [4]. Hence, Equation (2.43) implies

$$\vec{\nabla} \cdot (\rho_0(1 + k)\vec{v}) = -\frac{\partial \rho_0(1 + k)}{\partial t} \tag{2.44}$$

Since k is small, $\vec{\nabla} \cdot k\vec{v}$ can be neglected, and thus, we obtain by simplification

$$\vec{\nabla} \cdot \vec{v} = -\frac{\partial k}{\partial t} \tag{2.45}$$

Using Equation (2.43) in (2.45), we get

$$\nabla^2 \phi = \frac{\partial k}{\partial t} \tag{2.46}$$

Using $\vec{F} = 0$ and neglecting $(\vec{v} \cdot \vec{\nabla})\vec{v}$, Equation (2.42) implies

$$\frac{\partial \vec{v}}{\partial t} = -\frac{1}{\rho}\vec{\nabla}p \tag{2.47}$$

Using Equation (2.43), we get

$$\frac{\partial(\vec{\nabla}\phi)}{\partial t} = \frac{1}{\rho}\vec{\nabla}p \tag{2.48}$$

Multiply by $d\vec{r}$ [displacement between two points (x,y,z) and $(x+dx,y+dy,z+dz)$] on both sides to obtain

$$\frac{\partial(\vec{\nabla}\phi \cdot d\vec{r})}{\partial t} = \frac{1}{\rho}\vec{\nabla}p \cdot d\vec{r} \tag{2.49}$$

Hence, , $\vec{\nabla}\phi \cdot d\vec{r} = d\phi$ and $\vec{\nabla}p \cdot d\vec{r} = dp$. Thus, Equation (2.49) can be written as

$$\frac{\partial(d\phi)}{\partial t} = \frac{dp}{\rho} \tag{2.50}$$

Integrate Equation (2.50) to obtain

$$\frac{\partial(\phi)}{\partial t} = \int \frac{dp}{\rho} + C \tag{2.51}$$

Laplace stated that the process of sound propagation through air is an adiabatic process, being a fast process [1]. Thus, using adiabatic gas equation $pV\gamma = $ constant. Here, γ is called adiabatic gas constant, and its value is 1.41 for air. Using $m = \rho V$, where m is the mass of the gas, we get

$$p\left(\frac{m}{\rho}\right)\gamma = \text{constant}$$

Or, $p = S\rho^{\gamma}$, where S is the constant. Differentiating this, and using $\rho = \rho_0(1+k)$, we obtain

$$dp = S\gamma\rho_0{}^{\gamma}(1+k)^{\gamma-1}dk \tag{2.52}$$

Using Equation (2.52) in (2.51) we get

$$\frac{\partial \phi}{\partial t} = \int \frac{S\gamma\rho_0^{\gamma}(1+k)^{\gamma-1}dk}{\rho_0(1+k)} + C \tag{2.53}$$

Differentiate Equation (2.53) partially with respect to t

$$\frac{\partial^2 \phi}{\partial t^2} = S\gamma\rho_0^{\gamma-1}(1+k)^{\gamma-2}\frac{\partial k}{\partial t} \tag{2.54}$$

As stated earlier, k is very small for a gas. So, $(1+k)^{\gamma-2}$ is close to unity. Now, we can write

$$S\rho_0^{\gamma-1} = \frac{S\rho_0^{\gamma}}{\rho_0} = \frac{p_0}{\rho_0}$$

Thus, Equation (2.54) becomes

$$\frac{\partial^2 \phi}{\partial t^2} = \gamma \frac{p_0}{\rho_0}\frac{\partial k}{\partial t} \tag{2.55}$$

Using Equation (2.46) in (2.55), we get

$$\frac{\partial^2 \phi}{\partial t^2} = \gamma \frac{p_0}{\rho_0} \nabla^2 \phi$$

$$\nabla^2 \phi = \frac{\rho_0}{\gamma p_0} \frac{\partial^2 \phi}{\partial t^2} \tag{2.56}$$

Comparing this with the standard equation of a wave, $\nabla^2 y = \frac{1}{c^2}\frac{\partial^2 \phi}{\partial t^2}$, c is the velocity of wave. We get $c^2 = \gamma \frac{p_0}{\rho_0}$. So, $c = \sqrt{\frac{\gamma p_0}{\rho_0}}$. This is the expression for the propagation of wave.

2.4 DISCUSSION AND CONCLUSIONS

It can be concluded that differential equations play a pivotal role in solving physical systems. The partial differential equations have been implied to find the equation of a wave on a vibrating string and to find the speed of sound propagating in air. Similarly, partial differential equation method can also be used to study several other problems. The method of partial differential equations can be applied to solve problems like variable heat flow in a rod, two-dimensional and three-dimensional fluid flow, and propagation of waves in waveguides/transmission lines [1].

REFERENCES

1. Prakash, S. (2011). *Mathematical Physics Including Classical Mechanics*. Sultan Chand & Sons, New Delhi, India.
2. Kirkwood, J. (2019). *Mathematical Physics with Partial Differential Equations*. Academic Press, Cambridge, MA.
3. Strauss, W. A. (2007). *Partial Differential Equations an Introduction*. John Wiley & Sons, Hoboken, NJ.
4. Hsu, T. R. (2018). *Applied Engineering Analysis*. John Wiley & Sons, Hoboken, NJ.

3 Solution to One-Dimensional Heat Equation with Nonlocal Boundary Condition

L. Jones Tarcius Doss and S. Yaazhini

3.1 INTRODUCTION

In this chapter, a one-dimensional parabolic partial differential equation subjected to nonlocal boundary conditions is considered as the model problem:

$$u_t = u_{xx} + f(x,t) \qquad \text{where} \quad (x,y) \in [0,1] \times [0,T] \quad \text{and } t > 0 \tag{3.1}$$

with initial condition:

$$u(x,0) = g(x) \qquad \text{where} \quad x \in [0,1] \tag{3.2}$$

with boundary conditions:

$$u(0,t) = \int_0^1 \alpha_0 u(x,t)dx + g_0(t), \ t \in [0,T] \tag{3.3}$$

$$u(1,t) = \int_0^1 \alpha_1 u(x,t)dx + g_1(t), \ t \in [0,T] \tag{3.4}$$

In the past few decades, the nonlocal initial and boundary value problems have become a rapidly growing area of research. Many such models occur in the disciplines of engineering and life sciences where these are extracted in the form of partial differential equations with nonlocal initial and boundary conditions. There are several problems that arise in heat conduction [6], plasma physics [21], inverse problems [24], and thermodynamics [8]. In Ref. [14], the existence and uniqueness for a class of nonlocal nonlinear parabolic boundary value problems are derived, and the finite difference methods like backward Euler and Crank Nicolson methods were studied to support the theoretical findings. The one-dimensional heat equation with nonlocal initial condition was investigated in Ref. [18] by conversion to Fredholm integral equation. The semi-discretization approach was used to convert these partial differential equations with nonlocal boundary conditions into a system of first-order linear differential equations, and the equations were tested numerically according to Ref. [9]. The nonlocal boundary value problem in Laplace transform domain was studied in Ref. [2] where the problem was reduced to boundary value problems governed by second-order inhomogeneous ordinary differential equations that can be solved explicitly. The physical solution was then recovered by using numerical techniques if the inversion of the solution in the Laplace transform domain cannot be carried out exactly.

The analytical and approximate solutions were obtained in rapidly convergent series with computable components by Adomain decomposition technique for 2D time-dependent diffusion equation with the nonlocal boundary conditions in Ref. [10]. The existence property of boundary value problem for higher-order ordinary differential equations was derived in Ref. [25]. The nonlocal boundary value conditions of partial differential equations were solved in Ref. [13] by the method of lines, and numerical experiments were carried out in support of this theory. Crandall's method

DOI: 10.1201/9781003407386-3

"

to solve nonlocal boundary value problems was employed in Ref. [16] and was proved to have fourth-order convergence.

In Ref. [5], the nonlocal initial and boundary value problems of the integral type for the linear and nonlinear parabolic and hyperbolic partial differential equations were solved after converting into local Dirichlet boundary value problem. Analysis of the Crank Nicolson Hermite cubic orthogonal spline collocation method for nonlocal boundary value problems was performed in Ref. [3]. The matrix formulated algorithm was used to solve nonlinear reaction diffusion equations with nonlocal boundary conditions in Ref. [4]. The existence, uniqueness, and numerical solution of the parabolic interodifferential equations with nonlocal boundary conditions were studied in Ref. [22]. Parabolic and hyperbolic partial differential equations with nonlocal boundary conditions were solved using restarted Adomain decomposition method(RADM), which was used to convert the nonlocal boundary condition. In Ref. [17], the nonlocal initial value problem of one-dimensional heat equation was studied using the two-level finite difference schemes and found that at particular cases where the size of the time step is so small, the FTCS schemes are stable over the implicit schemes. In Ref. [7], various finite difference schemes are there to solve nonlinear nonlocal boundary conditions, and those methods are found to have acceptable rate of convergence [23]. The implicit Crandall's scheme (ICS) and the explicit Crandall's scheme (ECS) are used for solving one-dimensional heat equation with nonlinear nonlocal boundary value problems, and ICS requires more computational time over ECS, where the ECS is restricted for $r = 1/6$.

This chapter is classified as follows: Crank Nicolson with Weddle's quadrature rule is considered for finding the solution for the nonclassical model problem in the next session. In order to verify the theoretical results, numerical experiments are carried out, and the results are provided in the form of tables or figures in Section 3.3. Error analysis of Simpson's 1/3rd rule and the quadrature rule is provided in Section 3.4. Section 3.5 gives the brief conclusion using the results drawn from Sections 3.3 and 3.4.

3.2 CRANK NICOLSON WITH WEDDLE'S QUADRATURE RULE

We discretize the spatial domain of the Equation (3.1) into an $N \times N$ matrix a with step size $h = 1/N$ and the time step $k = T/M \cdot (x_i, t_j)$ are the grid points that are given by

$$x_i = ih, i = 0, 1, 2, \ldots, N \tag{3.5}$$

$$t_j = jk, j = 0, 1, 2, \ldots, M \tag{3.6}$$

where N should be a multiple of 6. u_i^j denoted the approximation of the exact solution at the node (i, j). The solution vector U^j is as given below:

$$U^j = (u_0^j, u_1^j, \ldots, u_N^j) \tag{3.7}$$

We use an implicit numerical method, namely, the Crank Nicolson scheme, as the foundation of our technique in Equation (3.1). This scheme is obtained by replacing the derivative terms by difference operators. Forward difference is applied along the time variable and average of central difference approximation at t_j, and t_{j+1} is applied along the space variable [12].

$$(\Delta_t u_i^j) = \frac{r}{2} \left[\delta_x^2 u_i^j + \delta_x^2 u_i^{j+1} \right] + \frac{k}{2} \left(f_i^{j+1} + f_i^j \right) \tag{3.8}$$

where $r = \frac{k}{h^2}$, $\Delta_t u_i^j = u_i^{j+1} - u_i^j$, and $\delta_x^2 u_i^j = u_{i-1}^j - 2u_i^j + u_{i+1}^j$.

After few rearrangements of the above scheme, the following formula for finding the solution of the heat equation is obtained

$$-ru_{i-1}^{j+1} + 2(1+r)u_i^{j+1} - ru_{i+1}^{j+1} = ru_{i-1}^{j} + 2(1-r)u_i^{j} + ru_{i+1}^{j} + \frac{k}{2}(f_i^{j+1} + f_i^{j}) \qquad (3.9)$$

for $\quad i = 1, 2, 3, \ldots, N-1 \quad$ and $\quad j = 0, 1, 2, \ldots, N$.

Equation (3.9) gives $N-1$ linear equations in $N+1$ unknowns $u_0, u_1, \ldots, u_N$. In order to solve the system of linear algebraic equations described above, we need two more equations. The integral boundary conditions defined in Equations (3.3) and (3.4) can be approximated to finite summation by Weddle's quadrature rule [20].

$$u_0^{j+1} = u(0, (j+1)k) = \int_0^1 \alpha_0(x) u(x, t^{j+1}) dx + g_0(t^{j+1})$$

$$\approx \frac{3h}{10}[(\alpha_0(x_0)u_0^{j+1} + 2(\alpha_0(x_6)u_6^{j+1} + \alpha_0(x_{12})u_{12}^{j+1} + \cdots + \alpha_0(x_{N-6})u_{N-6}^{j+1}) + 5(\alpha_0(x_1)u_1^{j+1}$$

$$+ \alpha_0(x_7)u_7^{j+1} + \cdots + \alpha_0(x_{N-1})u_{N-1}^{j+1}) + 6(\alpha_0(x_3)u_3^{j+1} + \alpha_0(x_9)u_9^{j+1} + \cdots + \alpha_0(x_{N-3})u_{N-3}^{j+1}) +$$

$$+ \alpha_0(x_N)u_N^{j+1}] + g_0(t^{j+1})$$

$$(3.10)$$

$$u_N^{j+1} = u(Nh, (j+1)k) = \int_0^1 \alpha_1(x) u(x, t^{j+1}) dx + g_1(t^{j+1})$$

$$\approx \frac{3h}{10}[(\alpha_1(x_0)u_0^{j+1} + 2(\alpha_1(x_6)u_6^{j+1} + \alpha_1(x_{12})u_{12}^{j+1} + \cdots + \alpha_1(x_{N-6})u_{N-6}^{j+1}) + 5(\alpha_1(x_1)u_1^{j+1}$$

$$+ \alpha_1(x_7)u_7^{j+1} + \cdots + \alpha_1(x_{N-1})u_{N-1}^{j+1}) + 6(\alpha_1(x_3)u_3^{j+1} + \alpha_1(x_9)u_9^{j+1} + \cdots + \alpha_1(x_{N-3})u_{N-3}^{j+1}) +$$

$$+ \alpha_1(x_N)u_N^{j+1}] + g_1(t^{j+1})$$

$$(3.11)$$

Rearranging the above terms of the Equations (3.10) and (3.11), the discretized boundary conditions are obtained as given below:

$$[\underbrace{(3h\alpha_0 x_0 - 10)}_{A_0}u_0^{j+1} + \underbrace{15h\alpha_0 x_1}_{A_1}u_1^{j+1} + \cdots + \underbrace{15h\alpha_0 x_{N-1}}_{A_{N-1}}u_{N-1}^{j+1} + \underbrace{3h\alpha_0 x_N}_{A_N}u_N^{j+1}] = -10g_0(t^{j+1})$$

$$(3.12)$$

$$[\underbrace{3h\alpha_1 x_0}_{B_0}u_0^{j+1} + \underbrace{15h\alpha_1 x_1}_{B_1}u_1^{j+1} + \cdots + \underbrace{15h\alpha_1 x_{N-1}}_{B_{N-1}}u_{N-1}^{j+1} + \underbrace{(3h\alpha_1 x_N - 10)}_{B_N}u_N^{j+1}] = -10g_1(t^{j+1})$$

$$(3.13)$$

The value of Equation (3.1) at the points $i = 0$ and $i = N$ is obtained from the set of Equations (3.12) and (3.13). The matrix form of the discretization of heat equation with nonlocal boundary conditions is obtained as follows. This is a $(N+1) \times (N+1)$ matrix

$$
\underbrace{\begin{bmatrix}
A_0 & A_1 & \cdots & \cdots & \cdots & \cdots & A_{N-1} & A_N \\
-r & (2+2r) & -r & 0 & \cdots & \cdots & 0 & 0 \\
0 & -r & (2+2r) & -r & 0 & \cdots & \cdots & 0 \\
\vdots & 0 & \ddots & \ddots & \ddots & & \vdots & \vdots \\
\vdots & \vdots & & \ddots & \ddots & \ddots & 0 & \vdots \\
0 & \vdots & \cdots & 0 & -r & (2+2r) & -r & 0 \\
0 & 0 & \cdots & \cdots & 0 & -r & (2+2r) & -r \\
B_0 & B_1 & & & & & B_{N-1} & B_N
\end{bmatrix}}_{L}
\times
\underbrace{\begin{bmatrix}
u_0^{j+1} \\ u_1^{j+1} \\ \vdots \\ \vdots \\ \vdots \\ \vdots \\ u_{N-1}^{j+1} \\ u_N^{j+1}
\end{bmatrix}}_{U^{j+1}}
=
$$

$$
\underbrace{\begin{bmatrix}
A_0 & A_1 & \cdots & \cdots & \cdots & \cdots & A_{N-1} & A_N \\
r & (2-2r) & r & 0 & \cdots & \cdots & 0 & 0 \\
0 & r & (2-2r) & r & 0 & \cdots & \cdots & 0 \\
\vdots & 0 & \ddots & \ddots & \ddots & & \vdots & \vdots \\
\vdots & \vdots & & \ddots & \ddots & \ddots & 0 & \vdots \\
0 & \vdots & \cdots & 0 & r & (2-2r) & r & 0 \\
0 & 0 & \cdots & \cdots & 0 & r & (2-2r) & r \\
B_0 & B_1 & & & & & B_{N-1} & B_N
\end{bmatrix}}_{M}
\times
\underbrace{\begin{bmatrix}
u_0^{j} \\ u_1^{j} \\ \vdots \\ \vdots \\ \vdots \\ \vdots \\ u_{N-1}^{j} \\ u_N^{j}
\end{bmatrix}}_{U^{j}}
+\frac{k}{2}
\underbrace{\begin{bmatrix}
f_0^{j+1}+f_0^{j} \\ \vdots \\ \vdots \\ \vdots \\ \vdots \\ \vdots \\ \vdots \\ f_N^{j+1}+f_N^{j}
\end{bmatrix}}_{\hat{f}}
$$

$$(3.14)$$

where the first and last row entries of the matrix R are zeroes for the linear nonlocal boundary conditions. The system of algebraic equations thus formed from the matrix in Equation (3.14) is represented as

$$Lu^{j+1} = Mu^{j} + \hat{f} \tag{3.15}$$

Observe that all the terms in the right hand side of the Equation (3.15) are all known, and these terms are represented together by the term $\hat{e}$

$$
\begin{bmatrix}
A_0 & A_1 & \cdots & \cdots & \cdots & \cdots & A_{N-1} & A_N \\
-r & (2+2r) & -r & 0 & \cdots & \cdots & 0 & 0 \\
0 & -r & (2+2r) & -r & 0 & \cdots & \cdots & 0 \\
\vdots & 0 & \ddots & \ddots & \ddots & & \vdots & \vdots \\
\vdots & \vdots & & \ddots & \ddots & \ddots & 0 & \vdots \\
0 & \vdots & \cdots & 0 & -r & (2+2r) & -r & 0 \\
0 & 0 & \cdots & \cdots & 0 & -r & (2+2r) & -r \\
B_0 & B_1 & & & & & B_{N-1} & B_N
\end{bmatrix}
\times
\begin{bmatrix}
u_0^{j+1} \\ u_1^{j+1} \\ \vdots \\ \vdots \\ \vdots \\ \vdots \\ u_{N-1}^{j+1} \\ u_N^{j+1}
\end{bmatrix}
=
\underbrace{\begin{bmatrix}
a_0^{j} \\ a_1^{j} \\ \vdots \\ \vdots \\ \vdots \\ \vdots \\ a_{N-1}^{j} \\ a_N^{j}
\end{bmatrix}}_{\hat{e}}
\tag{3.16}
$$

where $\hat{e} = Mu^{n} + \hat{f}$

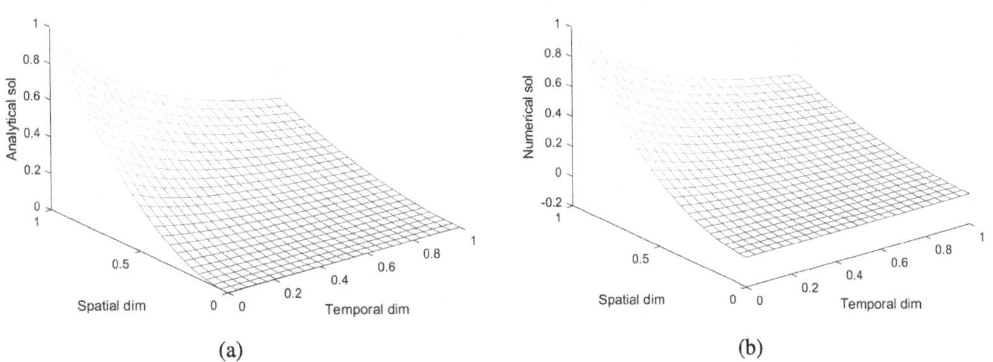

Figure 3.1 Exact and numerical solution of the test problem 1 by CN-Weddle's method. (a) Exact solution and (b) numerical solution.

3.3 NUMERICAL EXAMPLES

In this section, numerical examples are presented, and they are used to show the difference in relative error of the Simpson's 1/3rd and the Weddle's quadrature rule. The order of convergence of the proposed method is also tabulated (Figure 3.1).

3.3.1 TEST PROBLEM 1

Test Problem: 1 [23]

$$u_t - u_{xx} = \frac{-2(x^2 + t + 1)}{(t+1)^3}, \Omega = [0,1] \times [0,T]$$

Initial condition:

$$u(x,0) = x^2, x \in [0,1]$$

Boundary conditions:

$$u(0,t) = \int_0^1 x\,dx + \left(\frac{-1}{4(t+1)^2}\right), t \in [0,T]$$

$$u(1,t) = \int_0^1 x\,dx + \left(\frac{3}{4(t+1)^2}\right), t \in [0,T].$$

Exact solution

$$u(x,t) = \left(\frac{x}{t+1}\right)^2$$

The L_2 error at the ith spatial node between the numerical solution u_i and the exact solution U_i, is given by Equation (3.17), where $i = 0, 1, 2, \ldots, N$, and N represents the total number of grid points. The rate of convergence for this method for u using the L_2 norm is given by Equation (3.18).

$$||e_i|| = \sqrt{\sum_{i=0}^{N} \frac{(u_i - U_i)^2}{N}} \tag{3.17}$$

Table 3.1

Relative Error for $u(0.5,1)$ for Test Problem 1

Spatial Length	Simpson's 1/3rd Rule	Weddle's Rule
h=0.16667	2.496×10^{-1}	4.0864×10^{-3}
h=0.08333	1.248×10^{-1}	8.9763×10^{-4}
h=0.04166	6.240×10^{-2}	2.2624×10^{-4}

$$r = \left(\frac{\log \frac{||u_{h_1} - U||_2}{||u_{h_2} - U||_2}}{\log \frac{h_1}{h_2}} \right) \tag{3.18}$$

The L_∞ error between the numerical solution u_i at the ith spatial node and the exact solution U_i at the ith spatial node is given by Equation (3.19), and the rate of convergence for this method for u using the L_2 norm is given by Equation (3.20).

$$||e_i|| = \max_{x(i) \in [0,\infty]} |u_i - U_i| \tag{3.19}$$

$e_i's$ are vectors along the fixed time.

$$r = \left(\frac{\log \frac{||u_{h_1} - U||_\infty}{||u_{h_2} - U||_\infty}}{\log \frac{h_1}{h_2}} \right) \tag{3.20}$$

where h_1 and h_2 represent various step sizes, and u_{h_1} and u_{h_2} represent the numerical solution at h_1 and h_2, respectively.

The comparison of Simpson's 1/3rd and Weddle's quadrature rule (Table 3.1) is done using relative errors [16] given by

$$\text{Relative error} = \frac{|u - U_{\text{exact}}|}{|U_{\text{exact}}|} \tag{3.21}$$

3.3.1.1 Tables

Table 3.2 shows the numerical results for the relative error at $u(0.5, 1)$ using both the Simpson's $\frac{1}{3}$ and Weddle's quadrature rule.

Table 3.2

Convergence Analysis of Test Problem 1 Using L_2 and L_∞ Error Estimates at $t=0.5$

Spatial Grid Size	$\|u_h - U\|_2$	Rate of Convergence in L_2 Norm	$\|u_h - U\|_\infty$	Rate of Convergence in L_∞ Norm
$h = 0.16667$	7.5283×10^{-4}	2.03801	9.0120×10^{-4}	2.00984
$h = 0.08333$	1.8334×10^{-4}	2.00910	2.2380×10^{-4}	2.00855
$h = 0.04166$	4.5553×10^{-5}	2.00442	5.5627×10^{-5}	2.00855
$h = 0.02083$	1.1355×10^{-5}	2.00221	1.3898×10^{-5}	2.00044
$h = 0.010416$	2.8348×10^{-6}		3.4739×10^{-6}	

3.3.2 TEST PROBLEM 2

Test Problem: 2 [16]

$$u_t - u_{xx} = -e^{-(x+\sin t)}(1+\cos t), \quad \Omega = [0,1] \times [0,T]$$

Initial condition:

$$u(x,0) = e^{-x}, \quad x \in [0,1]$$

Nonlocal boundary conditions:

$$u(0,t) = \int_0^1 \left(\frac{e}{e-2} \right) x\,dx, \quad t \in [0,T]$$

$$u(1,t) = \int_0^1 \left(\frac{2}{e + \sin 1 - \cos 1} \right) \cos x\,dx, \quad t \in [0,T].$$

Exact solution

$$u(x,t) = e^{-(x+\sin t)}$$

The L_2 and L_∞ errors are computed in Equations (3.17)–(3.19) and are tabulated. The relative error of the Simpson's 1/3rd and Weddle's quadrature rule is found and displayed in Table 3.3 using the formula in Equation (3.21). It is evident from the tabular columns that Weddle's quadrature rule has better approximation at each spatial step size and is more consistent than the other method.

Table 3.4 shows the numerical results for the relative error at u(0.5,1) using both the Simpson's 1/3rd and Weddle's quadrature rule (Figure 3.2).

Table 3.3

Relative Error for $u(0.5,1)$ for Test Problem 2

Spatial Length	Simpson's 1/3rd Rule	Weddle's Rule
$h = 0.16667$	4.829×10^{-1}	1.96015×10^{-3}
$h = 0.08333$	2.4091×10^{-1}	4.9659×10^{-4}
$h = 0.04166$	1.2007×10^{-1}	1.2721×10^{-4}

Table 3.4

Convergence Analysis of Test Problem 2 Using l_2 and l^∞ Error Estimates at $t=0.5$

Spatial Grid Size	$\|u_h - U\|_2$	Rate of Convergence in L_2 Norm	$\|u_h - U\|_\infty$	Rate of Convergence in L_∞ Norm
$h = 0.16667$	4.1198×10^{-4}	2.05341	5.6533×10^{-4}	1.98076
$h = 0.08333$	9.9266×10^{-5}	2.01255	1.4325×10^{-4}	1.97292
$h = 0.04166$	2.4605×10^{-5}	1.93048	3.6496×10^{-5}	1.90348
$h = 0.02083$	6.4558×10^{-6}	1.69937	9.7566×10^{-6}	1.66629
$h = 0.010416$	1.9881×10^{-6}		3.0743×10^{-6}	

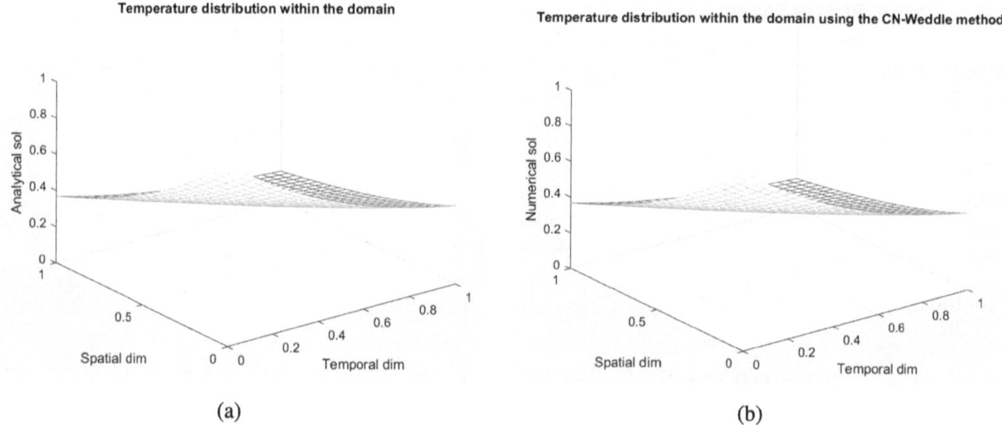

Figure 3.2 Exact and numerical solution of the test problem 1 by CN-Weddle's method. (a) Exact solution and (b) Numerical solution.

3.4 STABILITY OF THE PROPOSED METHOD

In this section, we analyze the stability of the proposed method. The stability condition for the above problem is derived using Gershgorin's circle theorem [15].

The numerical and the computational solutions of the Equation (3.1) are denoted by U and u, respectively. The linear algebraic equation of the matrix form in Equation (3.14) is thus given by

$$Lu^{j+1} = Mu^j + \hat{f} \tag{3.22}$$

$$LU^{j+1} = MU^j + \hat{f} \tag{3.23}$$

where $j = 0, 1, 2, \ldots, N$.

$$Le^{j+1} = Me^j$$

$$e^{j+1} = \underbrace{L^{-1}M}_{C} e^j$$

Thus, we get the equation of the form

$$e^{j+1} = Ce^j$$

$$\implies e^{j+1} = C(Ce^{j-1})$$

$$\vdots$$

$$\vdots$$

In general,

$$e^{j+1} = C^{j+1}e^0$$

where e^0 is the round off error at the initial level.

$$\implies ||e^{j+1}|| = ||C^{j+1}||||e^0|| \tag{3.24}$$

This error is controlled when the eigen value of $C \leq 1$ in magnitude. But, the eigen values of matrix C

$$C = \frac{\text{Maximum eigen values of matrix } M}{\text{Minimum eigen values of matrix } L} \leq 1 \tag{3.25}$$

On the application of Gershgorin's circle theorem [15], the location of eigen values is given as follows. The minimum row and column sums of the matrices L and M in Equation (3.14) are found:

Case(i) For $h \geq k$ where h and k are spatial and time steps.

Minimum value of row and column sums of the matrix L is $|2 + 4r|$.

Hence, the bound for eigen value is $|\lambda| \leq 2 + 4r$.

$$-2 - 4r \leq \lambda \leq 2 + 4r \tag{3.26}$$

Case(ii) For $h < k$

Minimum value of row and column sums of matrix L is $A_0 + A_1 + \cdots + A_N$

Hence the bound for eigen value is $|\lambda| \leq |A_0| + |A_1| + \cdots + |A_N|$

$$-|A_0| - |A_1| - \cdots - |A_N| \leq \lambda \leq |A_0| + |A_1| + \cdots + |A_N| \tag{3.27}$$

Repeat the same procedure for matrix M. For any value of h and k, the minimum value of row and column sums is $2 + 4r$. Hence, the bound for eigen value is $|\lambda| \leq 2 + 4r$.

$$-2 - 4r \leq \lambda \leq 2 + 4r \tag{3.28}$$

We now apply Brauer's theorem [11] to find the eigen values more appropriately. This theorem is applied to the model problem as follows: From first row of the matrix L in Equation (3.14), we have:

$$|\lambda - A_0| \leq |A_1| + |A_2| + |A_3| + \cdots + |A_N| \leq ||\beta_0||_\infty$$

$$\implies -||\beta_0||_\infty \leq \lambda - A_0 \leq ||\beta_0||_\infty$$

$$\implies (A_0 - ||\beta_0||_\infty) \leq \lambda \leq (A_0 + ||\beta_0||_\infty)$$

Applying similar work on other rows of the matrix:

$$|\lambda - 2 - 2r| \leq 2r$$

$$\implies 2 \leq \lambda \leq 2 + 2r$$

From last row of the matrix L, we have:

$$|\lambda - B_0| \le |B_1| + |B_2| + |B_3| + \cdots + |B_N| \le ||\beta_1||_\infty$$

$$\implies -||\beta_1||_\infty \le \lambda - B_0 \le ||\beta_1||_\infty$$

$$\implies (B_0 - ||\beta_1||_\infty) \le \lambda \le (B_0 + ||\beta_1||_\infty)$$

where $||\beta_0||_\infty, ||\beta_1||_\infty$ does not exceed twice if the magnitude of the diagonal entries. We find that for $h \le k$, the minimum of eigenvalues of L is $2 - 4r$, and for $h > k$, the minimum of eigenvalues of L is $A_0 - ||\beta_0||_\infty$.

The eigenvalues of matrix M from the Equation (3.14) using Gershgorin's circle theorem is found similarly as above.

$$|\lambda - (2 + 2r)| \le 2r$$

$$\implies 2 \le \lambda \le 2 + 4r$$

where the maximum eigenvalue for matrix M is $2 + 4r$.

$$\lambda \le 2 + 4r \tag{3.29}$$

We observe the minimum eigenvalues of the matrix L from the Equations (3.26) and (3.27) and, from the discussions above, the maximum eigen values of the matrix M from the Equations (3.28) and (3.29). Substituting them in Equation (3.25), we observe that the system is stable as the eigenvalues of $C \le 1$.

3.5 TRUNCATION ERROR

The order of convergence of both Crank Nicolson and Weddle's quadrature rule will be estimated in the subsection. The matrix Equation (3.14) of the problem (3.1) is a combination of both the schemes. Thus, the truncation error of both the schemes is analyzed. The error analysis is carried out to see the accuracy of approximation of the methods used to solve the given model problem (3.1). The central difference approximation for u_t at $u(x_i, t^{j+1/2})$ is given by Ref. [19]:

$$u_t(x_i, t^{j+1/2}) = \frac{u_i^{j+1} - u_i^j}{k} \tag{3.30}$$

which is of the order $O(k^2)$. The central difference approximation of u_{xx} [19] at $u(x_i, t^{j+1/2})$ is as follows, which is also of the order $O(h^2)$.

$$u_{xx}(x_i, t^{j+1/2}) = \frac{1}{2}\left[\frac{u_{i+1}^j - 2u_i^j + u_{i-1}^j}{h^2} + \frac{u_{i+1}^{j+1} - 2u_i^{j+1} + u_{i-1}^{j+1}}{h^2} \right] \tag{3.31}$$

For $j = 0, 1, 2, \ldots$ and for $i = 0, 1, 2, \ldots, N$, equating (3.30) and (3.31), we get

$$\frac{u_i^{j+1} - u_i^j}{k} = \frac{1}{2}\left[\frac{u_{i+1}^j - 2u_i^j + u_{i-1}^j}{h^2} + \frac{u_{i+1}^{j+1} - 2u_i^{j+1} + u_{i-1}^{j+1}}{h^2} \right] \tag{3.32}$$

Expanding each term of Equation (3.32) using the Taylor series at the point $(x_i, t^{j+\frac{1}{2}})$, we get

$$u(x_i, t^{j+1}) = u(x_i, t^{j+\frac{1}{2}}) + \left(\frac{k}{2}\right) u_t(x_i, t^{j+\frac{1}{2}}) + \left(\frac{k}{2}\right)^2 \frac{u_{tt}}{2!}(x_i, t^{j+\frac{1}{2}})$$

$$+ \left(\frac{k}{2}\right)^3 \frac{u_{ttt}}{3!}(x_i, \theta_1) + \cdots \quad \text{where} \quad \theta_1 \in (t^{j+\frac{1}{2}}, t^{j+1}) \tag{3.33}$$

$$u(x_i,t^j) = u(x_i,t^{j+\frac{1}{2}}) - \left(\frac{k}{2}\right)u_t(x_i,t^{j+\frac{1}{2}}) + \left(\frac{k}{2}\right)^2\frac{u_{tt}}{2!}(x_i,t^{j+\frac{1}{2}})$$

$$- \left(\frac{k}{2}\right)^3\frac{u_{ttt}}{3!}(x_i,\theta_1) + \cdots \qquad \text{where} \quad \theta_2 \in (t^j,t^{j+\frac{1}{2}}) \tag{3.34}$$

Subtracting Equation (3.34) from (3.33), we observe

$$u(x_i,t^{j+1}) - u(x_i,t^j) = ku_t(x_i,t^{j+\frac{1}{2}}) + \frac{k^3}{24}u_{ttt}(x_i,\theta) + \cdots \tag{3.35}$$

$$\text{where} \quad \theta \in (t^{j+\frac{1}{2}},t^{j+1})$$

$$u_t = \frac{u(x_i,t^{j+1}) - u(x_i,t^j)}{k} - \frac{k^2}{24}u_{ttt}(x_i,\theta) + \cdots \tag{3.36}$$

The second-order differentiation term u_{xx} is approximated by the average of the central difference method at the time level t^j and t^{j+1}

$$u(x_i + h,t^j) = u(x_i,t^j) + hu_x(x_i,t^j) + \left(\frac{h^2}{2!}\right)u_{xx}(x_i,t^j) + \left(\frac{h^3}{3!}\right)u_{xxx}(x_i,t^j)$$

$$+ \left(\frac{h^4}{4!}\right)u_{xxxx}(\eta_1,t^j)\ldots \qquad \text{where} \quad \eta_1 \in (x_i,x_{i+1}) \tag{3.37}$$

$$u(x_i - h,t^j) = u(x_i,t^j) - hu_x(x_i,t^j) + \left(\frac{h^2}{2!}\right)u_{xx}(x_i,t^j) - \left(\frac{h^3}{3!}\right)u_{xxx}(x_i,t^j)$$

$$+ \left(\frac{h^4}{4!}\right)u_{xxxx}(\eta_2,t^j)\ldots \qquad \text{where} \quad \eta_2 \in (x_{i-1},x_i) \tag{3.38}$$

Adding the Equations (3.37) and (3.38) gives the following:

$$u(x_i + h,t^j) + u(x_i - h,t^j) = 2u(x_i,t^j) + h^2 u_{xx}(x_i,t^j) + \frac{h^4}{12}u_{xxxx}(\eta,t^j) \tag{3.39}$$

$$\text{where} \quad \eta \in (x_{i-1},x_{i+1})$$

$$u_{xx}(x_i,t^j) = \frac{u(x_i + h,t^j) - 2u(x_i,t^j) + u(x_i - h,t^j)}{h^2} - \frac{h^2}{12}u_{xxxx}(\eta,t^j) \tag{3.40}$$

Following the same procedure for the approximation of u_{xx} at the time level t^{j+1}

$$u_{xx}(x_i,t^{j+1}) = \frac{u(x_i + h,t^{j+1}) - 2u(x_i,t^{j+1}) + u(x_i - h,t^{j+1})}{h^2} - \frac{h^2}{12}u_{xxxx}(\eta,t^{j+1}) \tag{3.41}$$

Now taking the average of the Equations (3.40) and (3.41), we get

$$u_{xx} = \frac{1}{2}\left(\frac{u_{i+1}^j - 2u_i^j + u_{i-1}^j}{h^2} + \frac{u_{i+1}^{j+1} - 2u_i^{j+1} + u_{i-1}^{j+1}}{h^2}\right) - \frac{h^2}{12}u_{xxxx}(\eta,t^j) - \frac{h^2}{12}u_{xxxx}(\eta,t^{j+1}) \tag{3.42}$$

Adding Equations (3.36) and (3.42) and comparing the resultant equation with Equation (3.32), we get the order of convergence of the scheme to be of the order 2 for both time and space variables.

$$\frac{u_i^{j+1} - u_i^j}{k} = \frac{1}{2}\left[\frac{u_{i+1}^j - 2u_i^j + u_{i-1}^j}{h^2} + \frac{u_{i+1}^{j+1} - 2u_i^{j+1} + u_{i-1}^{j+1}}{h^2}\right]$$

$$- \frac{k^2}{24}u_{ttt} - \frac{h^2}{12}u_{xxxx} \tag{3.43}$$

$$\frac{u_i^{j+1} - u_i^j}{k} = \frac{1}{2} \left[\frac{u_{i+1}^j - 2u_i^j + u_{i-1}^j}{h^2} + \frac{u_{i+1}^{j+1} - 2u_i^{j+1} + u_{i-1}^{j+1}}{h^2} \right] + O(k^2) + O(h^2) \tag{3.44}$$

The order of convergence of the Weddle's rule can be derived from the error analysis of the Newton Cote's formula in Ref. [1] (which is a generalized form of all the quadrature rules). Suppose that $\sum_{i=0}^{N} a_i f(x_i)$ is the $n+1$ point that closed Newton cote's formula with $0 = x_0$ and $1 = x_n$ and $h = \frac{b-a}{N}$, there exists some $\psi \in (0,1)$ for which n is even and is given by

$$\int_0^1 f(x)dx = \sum_{i=0}^n a_i f(x_i) + \frac{h^{(n+3)} f^{(n+2)} \psi}{(n+2)!} \int_0^n t^2(t-1)(t-2)\dots(t-n)dt$$

where $x_i = 0 + ih$, $i = 0, 1, 2, \dots, n$.

Now in order to analyze the error involved in Weddle's rule which is employed in Equations (3.3) and (3.4), respectively, for $n = 6$ which is even,

$$E_6(f) = \frac{h^{(6+3)} f^{(6+2)}(\psi)}{(6+2)!} \int_0^6 t^2(t-1)(t-2),\dots(t-6)dt$$

$$= \frac{h^9 f^8(\psi)}{8!} \int_0^6 (t^8 - 21t^7 + 175t^6 - 735t^5 + 1,624t^4 - 1764t^3 + 720t^2)dt$$

The above equation is further integrated and simplified to get

$$E_6(f) = \frac{h^9 f^8(\psi)}{1,120} \left[\frac{-36}{5} \right] dt$$

$$= \frac{-9h^9 f^{(viii)}(\psi)}{1,400}$$

The magnitude of error in Weddle's rule is thus

$$|E_6(f)| \leq \frac{9h^9}{1,400} m_8$$

where $m_8 = \max_{\psi \in (0,1)} f^{(viii)}(\psi)$. Hence, we find that our model problem converges to the maximum order among the proposed methods.

3.6 CONCLUSION

In this chapter, a new method, namely, Crank Nicolson finite difference with Weddle's quadrature rule was employed to solve one-dimensional heat equations with nonlocal boundary conditions. Stability analysis and the truncation error of the proposed method were carried out and found that the method is stable for $|\lambda| < 1$, and the order of convergence is 2. The relative errors of Simpson's 1/3rd rule and Weddle's rule were found. The finite difference technique used in this chapter can be extended to higher dimensional heat equation with nonlocal boundary conditions.

REFERENCES

1. Al-Sammarraie, O.A. & Bashir, M.A. (2015). Error analysis of the high order Newton Cotes formulas, *International Journal of Scientific and Research Publications*, 5, 1–6.
2. Ang, W.T. (2003). A method of solution for the one-dimensional heat equation subject to nonlocal conditions, *Southeast Asian Bulletin of Mathematics*, 26, 185–191.
3. Bialecki, B., Fairweather, G. & Lopez-Marcos, J.C. (2013). The Crank-Nicolson hermite cubic orthogonal spline collocation method for the heat equation with nonlocal boundary conditions, *Advances in Applied Mathematics and Mechanics*, 5, 442–460.

4. Borhanifar, A., Shahmorad, S. & Feizi, E. (2017). A matrix formulated algorithm for solving parabolic equations with nonlocal boundary conditions, *Numerical Algorithms*, 74, 1203–1221.

5. Bougoffa, L. & Rach, R.C. (2013). Solving nonlocal initial-boundary value problems for linear and nonlinear parabolic and hyperbolic partial differential equations by the Adomian decomposition method, *Applied Mathematics and Computation*, 225, 50–61.

6. Bouziani, A. (1991). On a class of parabolic equationswith non local boundary conditions, *Academie Royale de Belgique. Bulletin de la Classe des Sciences*, 10, 61–67.

7. Bouziani, A., Bensaid, S. & Dehilis, S. (2020). A second order accurate difference scheme for the diffusion equation with nonlocal nonlinear boundary conditions, *Journal of Physical Mathematics*, 11, 1–6.

8. Day, W.A. (1992). Parabolic equation and thermodynamics, *Quarterly of Applied Mathematics*, 50, 523–533.

9. Dehghan, M. (2003). Numerical solution of a parabolic equation with non-local boundary specifications, *Applied Mathematics and Computation*, 145, 185–194.

10. Dehghan, M. (2004). Application of the Adomian decomposition method for two-dimensional parabolic equation subject to nonstandard boundary specifications, *Applied Mathematics and Computation*, 157, 549–560.

11. Harfash, A.J. & Jalob, H.A. (2006). Sixth and fourth order finite difference schemes for two and three dimension poisson equation with two methods to derive this schemes, *Basrah Journal of Science*, 24, 1–20.

12. Iyengar, S.R.K. & Jain, R.K. (2009). *Numerical Methods*. New Age International Publishers, New Delhi, pp. 281–283.

13. Javidi, M. (2006). The MOL solution for the one dimensional heat equation subject to nonlocal conditions, *International Mathematical Forum*, 1, 597–602.

14. Lin, Y. (1978). Analytical and numerical solutions for a class of nonlocal nonlinear parabolic differential equations, *Journal of Mathematical Analysis*, 271, 263–271.

15. Marquis, D. (2016). Gershgorin's Circle Theorem for estimating the eigenvalues of a matrix with known error bounds, pp. 6–8.

16. Martin-Vaquero, J. & Vigo-Aguiar, J. (2009). A note on efficient techniques for the second-order parabolic equation subject to non-local conditions, *Applied Numerical Mathematics*, 59, 1258–1264.

17. Martin-Vaquero, J. & Sajavicius, S. (2019). The two-level finite difference schemes for the heat equation with nonlocal initial condition, *Applied Mathematics and Computation*, 342, 166–177.

18. Olmstead, W.E. & Roberts, C.A. (1997). The one-dimensional heat equation with a nonlocal initial condition, *Applied Mathematics Letters*, 10, 89–94.

19. Peterson, J. (2003). Crank Nicholson scheme for the heat equation, *Applied Mathematics and Computation*, 145, 185–194.

20. Pollard, J.H. (1977). *A Handbook of Numerical and Statistical Techniques with Examples Mainly from the Life Sciences*, Cambridge University Press, Cambridge, UK, pp. 37–43.

21. Samarskii, A. (1980). Some problems in differential equation theory, *Differential Equations*, 16, 1221–1228.

22. Sofiane, D., Bouziani, A. & Taki Eddine, O. (2018). Study of solution for a parabolic integrodifferential equation with the second kind integral condition, *International Journal of Analysis and Applications*, 16, 569–593.

23. Souad, B., Sofiane, D. & Abdelfatah, B.(2021). Explicit and implicit Crandall's scheme for the heat equation with nonlocal nonlinear conditions, *International Journal of Analysis and Applications*, 19, 660–673.

24. Wang, S. & Lin, Y. (1989). A finite difference solution to an inverse problem for determining control functionin a parabolic partial differential equation, *Inverse Problems*, 5, 631–640.

25. Zengji, D., Xiaojie, L. & Weigao, G. (2006). Nonlocal boundary value problem of higher order ordinary differential equations at resonance, *Rocky Mountain Journal of Mathematics*, 36, 1471–1486.

4 Identification of Optimum Public EV Charging Stations

V. G. Deepa, S. Aparna Lakshmanan, and V. N. Sreeja

4.1 INTRODUCTION

Although transportation is a need of modern life, the ordinary internal combustion engine is gradually becoming outmoded. Currently, the majority of people cannot afford the rate at which fuel costs are rising. Roughly equivalent diesel or gasoline vehicles have far higher operating costs than an electric vehicle. These vehicles also produce a lot of pollutants. In this setting, completely electric vehicles (EVs) are becoming more and more relevant, and they are swiftly replacing other vehicle kinds. EVs don't emit any dangerous gases, making them substantially more environment friendly. Compared to gasoline or diesel automobiles, electric vehicles need less maintenance. As a result, running an electric automobile costs substantially less each year.

The ease of home charging is among the most often cited advantages of having an electric vehicle. This benefit is frequently applicable to people who have access to a power outlet adjacent or who have the capacity to charge at home using less expensive resources, such as solar power. The lack of charging stations is one of the main problems. Compared to gas, diesel, and petrol stations, there really are very few public EV charging stations, and the charging process takes longer. The electric vehicles need to recharge while they are moving if someone constantly drives outside. However, this is progressively less of a problem as even more power stations are put in place around the area. Regular people will benefit from the installation of electric vehicle charging stations at key sites nearby. The process of installing charging stations is very costly. It costs time in addition to money. The ideal number of stations that can be reached by the majority of users must thus be determined for each area. In the view of literature, it is see that Dong Ding et al. [1], studied electric vehicle charging warning and path planning method based on spark.

Graph theory is a major branch of mathematical science, which emerged as a result of search for the solution of the well-known "Konigsberg bridge problem". Oystein Ore first used the terminology "dominating set" and "domination number" in his book on graph theory in 1962, which led to the development of a strong branch of graph theory, called domination theory [3]. A set $D \subseteq V$ for the graph $G = (V, E)$ is a dominating set of G if every vertex, which is not included in D, is adjacent to at least one element of D. The cardinality of the smallest dominating set is known as domination number, $\gamma(G)$ of G. The chess problem, which is used to estimate the minimal number of queens needed on the chessboard so that all squares are either occupied or can be attacked by a queen, is associated with the historical foundations of domination [4]. The distance k-dominating set D is a subset of the vertex set $V(G)$ such that each $x \in V(G) - D$ is within a distance atmost k from some vertex in D. The k-domination number of G is the minimum cardinality over all distance k-dominating sets, which is denoted by $\gamma_k(G)$. For example, a graph G with $|G| = 12$, Figure 4.1 shows the dominating set of G in blue colored vertices. From this figure, we can see that the domination number of G is 4. Figure 4.2 shows the distance k-dominating set of G with $k = 2$ in blue colored vertices. Here, $\gamma_2(G) = 3$. Two two-dominating sets are included just to show that the k-dominating set need not be unique (Figures 4.2 and 4.3). Likewise, Figure 4.4 describes the distance k-dominating set with $k = 3$ in blue colored vertex and $\gamma_3(G) = 1$.

DOI: 10.1201/9781003407386-4

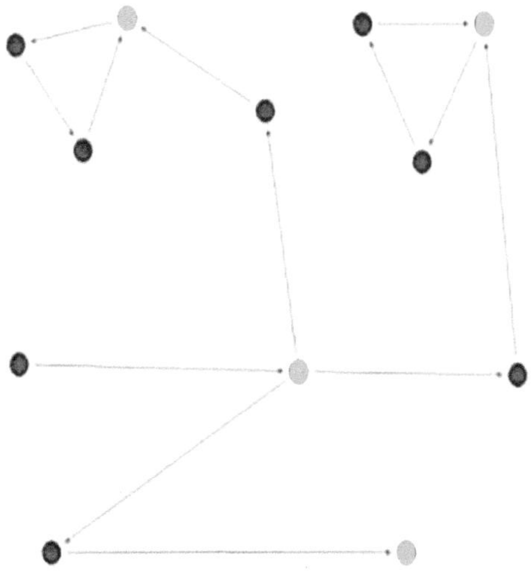

Figure 4.1 The dominating set of G.

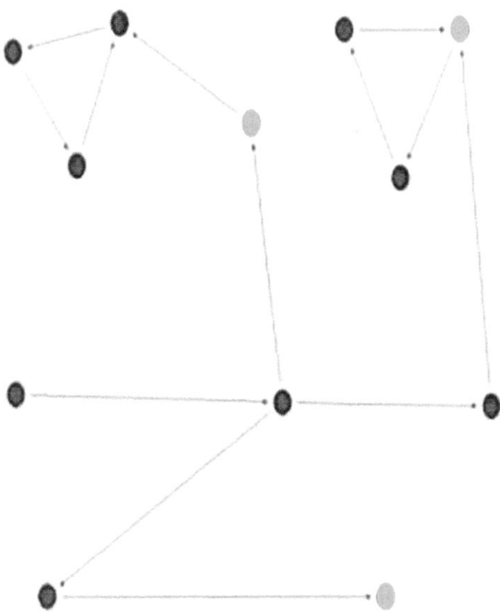

Figure 4.2 The distance two-dominating set of G.

4.1.1 OBJECTIVE OF THE STUDY

In this era, when electric vehicles are gaining popularity, there are not enough public charging stations is a big drawback. Installing a sufficient number of pricey charging stations quickly is not feasible. Therefore, the smallest number of charging stations should indeed be set up without impeding traffic. In this chapter, we tried to demonstrate that wherever charging stations are provided, people may travel smoothly. Of course, it is desirable to place charging stations next to each other on the road we travel on, but the heavy cost of installing each station, the space limitation, the labor cost, etc. are the main inferences. Therefore, it is important to estimate the location of the minimum

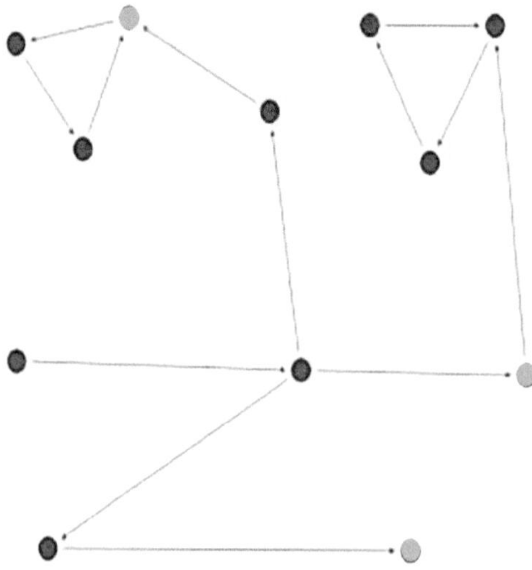

Figure 4.3 The distance two-dominating set of G.

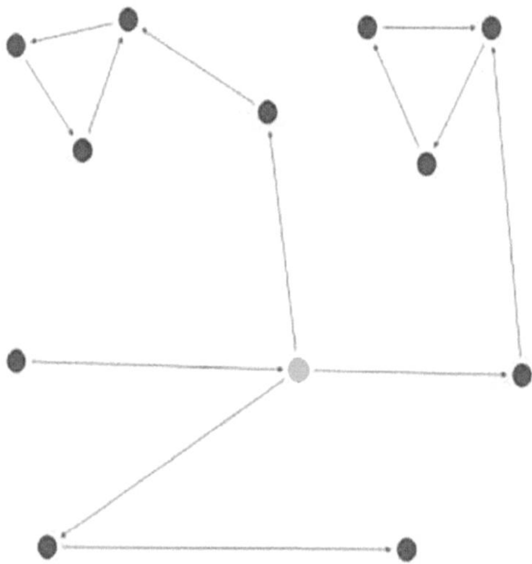

Figure 4.4 The distance three-dominating set of G.

stations that may be built while taking into account the smooth trip of every electric vehicle coming from every direction. We tried to solve this problem with the help of the famous k-dominating set problem in graph theory.

4.2 FORMATION OF ROADS AS A NETWORK

To find a logical solution for a real-life problem, it is easy to covert the problem as a mathematical one and find solution. For finding optimal charging stations, we used the methodology described below:

An ideal spot for placing charging stations is beside the road, so as to convert roads as a graph. For that, we took the vertex v_0 as the initial point of a road and the vertex v_1 is placed just 1 km

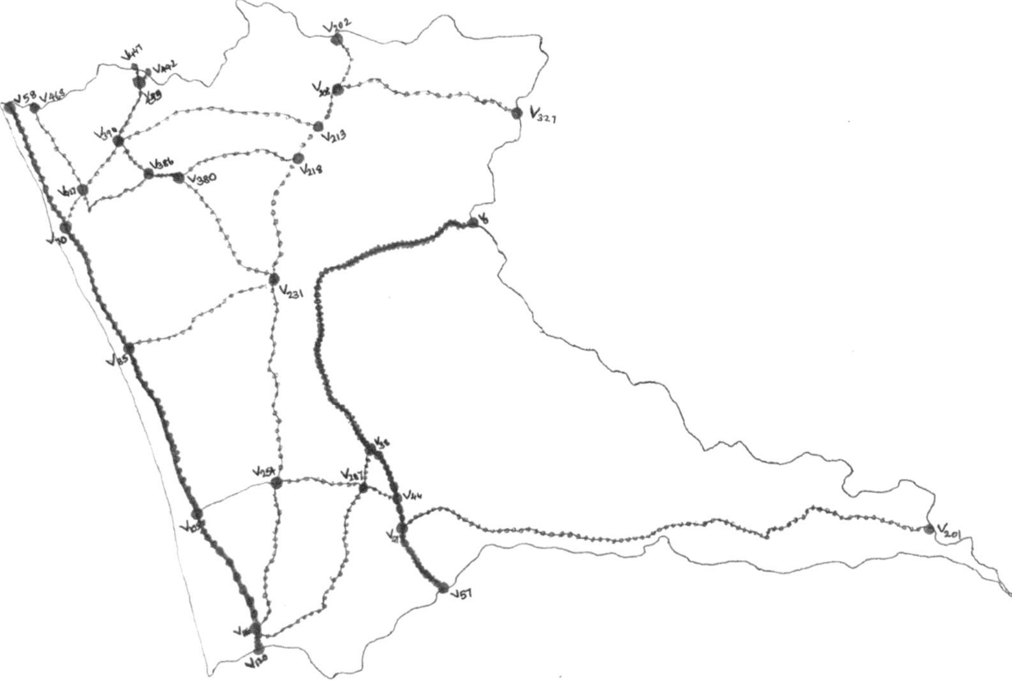

Figure 4.5   Road graph of Thrissur district, Kerala.

from the starting point. Similarly, v_2 is placed just 1 km from the vertex v_1. v_3 is from v_2 and so on. If the road has only n kilometers, the road ends at the vertex v_n. If the vertex v_r that comes in between v_0 and v_n is a junction, definitely $d(v_r) \geq 2$. Suppose the road ends at v_k; $0 \leq k < n$, then we take $v_{(k+1)}$ as the initial point of the next road.

We gathered information about the Thrissur district in Kerala State, India, as an example. Two major highways with a total length of 119 km and eight state highways with such a total length of 338 km travel through the district of Thrissur. As previously said, we created a network out of these roadways. If the distance is much <1 km, it might be disregarded. Since, just addressing one district, it goes without saying that both beginning and termination points for national roads (NH) exist. The initial point of NH 544 in Thrissur district is noted as v_0 and its terminal point is noted as v_{57}. The initial point of NH 66 in Thrissur district is marked as v_{58}, and this ends up in v_{120}. The state highway (SH 21) is from v_{44} and it ends in v_{201}. Similarly, all the national and state highways are marked up to v_{468}. The roads other than state and national highways are neglected. Clearly, the resultant is a connected graph, and it is given in Figure 4.5. For a quick reference between the road graph and original road structures, the road map of Thrissur district, Kerala, is given in Figure 4.6.

4.3   RELEVANCE OF DISTANCE K-DOMINATING SET IN SEARCH OF OPTIMUM CHARGING STATIONS

The primary issue while using an electric vehicle is what to do if the battery runs out while it is in motion. Nothing to worry about if there are enough public EV charging stations. However, this is an untrue fact. So, we move to some technical features of an electric vehicle. The EV will undoubtedly inform the driver when the charge is about to run out. There should be a public EV charging station inside k-kilometers of the car, either forward or backward, if the vehicle alerted the driver k-kilometres before the battery was about to run out of power. Everyone must choose

Figure 4.6   Road map of Thrissur district, Kerala.

forward for a comfortable ride, as we all know, but for a continuous route, an EV charging station may be found within k-kilometers without taking direction into account. Therefore, it must locate these locations. Because of how difficult this endeavor is in general, we used the idea of a distance k-dominated set. If it models the highways as previously indicated, the k-domination set of the graph is really the best location for the public EV charging stations. In other words, each vertex there in graph G is either an EV charging station that is open to the public or is within k-kilometer of one. Hence, the problem reduced as finding the k-domination set of the graph G.

The problem of determining a minimum dominating set is a classical one. Many of the mathematicians tried to solve this problem for a long time and found better upper bounds for the domination number [5]. In 2020, Minh Hai Nguyen et al. [2] solved the k-dominating set problem on very large-scale networks. They proposed an efficient heuristic algorithm that can handle up to "17 million vertices and 33 million edges". So, the only problem of finding optimal public EV charging stations is to model the required area as a graph by using the methodology which is described above. Hence, we can find the minimum k-dominating set by using Ref. [2].

In this illustration, we found dominating sets for Figure 4.5, with two different values of k. When $k = 20$, that is, if we get an alert from the vehicle just 20 km before the battery power become empty, it can reach to every point by placing charging stations in these eight vertices. That is, $\gamma_{20}(G) = 8$ and that one of the 20-dominating set of G is $\{v_{20}, v_{105}, v_{141}, v_{181}, v_{286}, v_{307}, v_{365}, v_{428}\}$. Similarly, when $k = 30$, that is, if we get an alert from the vehicle just 30 km before the battery power become empty, it can reach to every point by placing charging stations in these five vertices. That is, $\gamma_{30}(G) = 5$ and that one of the 30-dominating set of G is $\{v_{30}, v_{72}, v_{171}, v_{287}, v_{410}\}$. As we said earlier, domination set is not a unique one. Here, it explains only one solution of the problem. It is easy to find other suitable solutions too. However, k-domination number must be the same.

4.4   LIMITATIONS

The problem of converting road to a network is little bit time consuming. Since this is a practical problem, we need some logical solutions than a mathematical one. It is occasionally needed to abide by certain traffic laws to travel safely. In certain locations, vehicles only go in one direction (one-way traffic). This situation won't be taken into account when trying to locate a k-dominating set in an undirected graph. But by changing the value of k, it can get around this problem. Similar to this, heavy traffic block areas are not considered while finding k-dominating set.

4.5   CONCLUSION

When a scientific subject's applications spark movements among regular people as well, acceptance of that discipline rises. In this aspect, mathematics, particularly graph theory, is highly well-liked. Through this chapter, using the concept of dominance in graph theory, the authors have tried to define a very complicated problem. Although everyone is aware of the advantages of electric vehicles, a shortage of suitable public charging facilities is preventing more people from making the move. Taking advantage of domination's possibilities, it can be understood that wherever minimum charging stations are installed, seamless travel is possible. For an illustration, the authors found where the location of optimum charging stations is in Thrissur district of Kerala.

REFERENCES

1. Dong Ding, Junhuai Li, Pengjia Tu, Huaijun Wang, Ting Cao, and Facun Zhang (2020). Electric vehicle charging warning and path planning method based on spark, *IEEE Access*, 8. doi: 10.1109/ACCESS.2020.2964307.
2. Minh Hai Nguyen, Minh Hoang Ha, Diep N. Nguyen, and The Trung Tran (2020). Solving the k-dominating set problem on very large-scale networks, *Computational Social Networks*, 7: 4. https://doi.org/10.1186/s40649-020-00078-5.
3. Oystein Ore (1962). *Theory of Graphs*, American Mathematical Society Colloquium Publications, Providence, RI, Vol. 38. DOI: https://doi.org/10.1090/coll/038
4. Shobha Shukla and Vikas Singh thakur (2020). Domination and it's type in Graph theory, *JETIR*, 7(3): 1549–1557.
5. Stephen Travis Hedetniemi and Ravi Chandran Lascar (1990). Bibliography on domination in graphs and some basic definitions of domination parameters, *Discrete Mathematics*, 86(1–3): 257–277. doi: 10.1016/0012-365X(90)90365-O.

5 Spanning Soft Trees

Bobin George, Jinta Jose, and Rajesh K. Thumbakara

5.1 INTRODUCTION

In 1999, D. Molodtsov [11] introduced the revolutionary concept of soft set theory, a mathematical framework tailored to handle situations involving uncertainty. Authors such as R. Biswas, P. K. Maji, and A. R. Roy [9, 10] have employed soft set theory in various decision-making contexts. The inception of soft graphs is credited to R. K. Thumbakara and B. George [20], with further refinement of the concept undertaken by M. Akram and S. Nawaz [2] in 2015. Concurrently, M. Akram and F. Zafar [4] played a pivotal role in pioneering the concept of soft trees. M. Akram and S. Nawaz [2] introduced various operations in soft graphs, including the Cartesian and lexicographic products. In related work, J. Jose, B. George, and R. K. Thumbakara [7] conducted extensive research into the characteristics of Cartesian and lexicographic products of soft graphs.

In their publication, M. Akram and S. Nawaz [3] introduced diverse types of soft graphs, encompassing regular soft graphs, irregular soft graphs, neighborly irregular soft graphs, and highly irregular soft graphs, while investigating their properties. Additionally, they delved into the concept of cut vertices and bridges in soft graphs, elucidating their respective properties. In a separate study, J. D. Thenge, B. S. Reddy, and R. S. Jain [16] defined parameters like radius, diameter, center, and degree in soft graphs, presenting associated properties. Their work [17] involved the definition of connected soft graphs and the derivation of related results. They [18] also introduced the concept of adjacency and incidence matrices in soft graphs, yielding pertinent results.

In their collaborative research, B. George, J. Jose, and R. K. Thumbakara introduced intriguing concepts such as Eulerian [8] and Hamiltonian soft graphs [22], subdivision graphs, power graphs, and line graphs for soft graphs [21], in addition to examining various product operations applicable to soft graphs, including tensor products, strong products [5], co-normal products, and modular products [6]. J. D. Thenge, B. S. Reddy, and R. S. Jain [19] presented an application of soft graphs in decision-making by leveraging the adjacency matrix of soft graphs to design an algorithm.

M. A. Abbood, A. A. J. Al-Swidi, and A. A. Omran [1] explored soft graphs of various graph types, including null graphs, complete graphs, cycle graphs, bipartite graphs, star graphs, and wheel graphs. N. Sarala and K. Manju [13] introduced various types of soft graphs, such as soft bipartite graphs, soft complete bipartite graphs, soft regular graphs, soft two-regular graphs, and soft regular bipartite graphs, while investigating their practical applications. R. K. Thumbakara, B. George, and J. Jose [23] introduced the concept of isomorphism in soft graphs and explored tabular representations of isomorphic soft graphs. They also investigated the concept of the soft complement of a soft graph and self-complementary soft graphs, examining their key properties. K. Palani, T. Jones, and V. Maheswari [12] discussed a range of soft graphs derived from well-known graph types, including complete graphs, star graphs, complete bipartite graphs, crown graphs, comb graphs, friendship graphs, bistar graphs, and wheel graphs.

N. Sarala and K. Manju [14] introduced concepts of domination in soft graphs, including independent dominating sets, accurate dominating sets, connected dominating sets, total dominating sets, isolate dominating sets, and distance-2 dominating sets, while investigating their inherent properties and proposing an algorithm for minimal dominating sets in soft graphs. In their work [15], they introduced the concept of detour domination in soft graphs, along with related concepts such as detour distance, total detour distance, detour dominating sets, minimal detour dominating sets,

DOI: 10.1201/9781003407386-5

and upper detour dominating numbers, exploring their properties. Venkatraman, R. Helen, and C. Natarajan [25] introduced the novel parameter of the soft domination number and computed this parameter for soft graphs derived from well-established graphs. S. Venkatraman and R. Helen [24] presented findings on the domination number of soft graphs.

Spanning trees, as unique subgraphs of a graph, possess various essential characteristics. Typically, they serve the purpose of determining the most efficient path for connecting all nodes within the graph. Several typical applications include their utilization in civil network planning to create networks, implementing network routing protocols, as well as being employed in clustering, cluster analysis, handwriting recognition, and image segmentation. In this chapter, we introduce the concept of spanning soft trees and investigate their properties.

5.2 PRELIMINARIES OF SOFT GRAPH

M. Akram and S. Nawaz [2,3] defined a soft graph as follows: "Let $H^* = (R,D)$ be a simple graph and Z be any nonempty set. Let L be an arbitrary relation between elements of Z and elements of R. That is, $L \subseteq Z \times R$. A mapping $W : Z \to P(R)$ can be defined as $W(z) = \{y \in R : zRy\}$. Also define a mapping $Y : Z \to P(D)$ by $Y(z) = \{uv \in D : \{u,v\} \subseteq W(z)\}$. The pair (W,Z) is a soft set over R and the pair (Y,Z) is a soft set over D. Then, the four-tuple $H = (H^*,W,Y,Z)$ is called a soft graph if it satisfies the following conditions:

1. $H^* = (R,D)$ is a simple graph,
2. Z is a nonempty set of parameters,
3. (W,Z) is a soft set over R,
4. (Y,Z) is a soft set over D,
5. $(W(z),Y(z))$ is a subgraph of H^* for all $z \in Z$.

If we represent $(W(z),Y(z))$ by $J(z)$, then soft graph H is also given by $\{J(z) : z \in Z\}$. Then, $J(z)$ corresponding to a parameter z in Z is called a part of the soft graph H. Let $H = (H^*,W,Y,Z)$ be a soft graph and v be any vertex of the part $J(z)$ of H for some $z \in Z$. Then, the part degree of the vertex v in $J(z)$ denoted by $d(v)[J(z)]$ is the degree of the vertex v in that part $J(z)$. A soft graph H is called a soft tree if the part $J(z)$ of H is a tree for every $z \in Z$. H is called a complete soft graph if the part $J(z)$ of H is complete for every $z \in Z$. H is called a soft cycle if the part $J(z)$ of H is a cycle for every $z \in Z$".

5.3 SPANNING SOFT TREES

Definition 5.1 *Let $H^* = (R,D)$ be a simple graph and $H = (H^*,W,Y,Z)$ be a soft graph of H^* which is also given by $\{J(z) : z \in Z\}$. Let $J(z)$ be a part of H for some $z \in Z$. Then, a subgraph $S(z)$ of $J(z)$ is called a spanning subgraph of $J(z)$ if the vertex set of $S(z)$ is the same as $W(z)$.*

Definition 5.2 *Let $J(z)$ be a part of a soft graph $H = (H^*,W,Y,Z)$ for some $z \in Z$. Then, a spanning tree of $J(z)$ is a spanning subgraph of $J(z)$ that is a tree. A spanning tree of $J(z)$ is denoted by $T(z)$ for all $z \in Z$.*

Definition 5.3 *Let $H = (H^*,W,Y,Z)$ be a soft graph of a simple graph $H^* = (R,D)$ represented by $\{J(z) : z \in Z\}$. Then, a spanning soft tree of H is given by $\{T(z) : z \in Z\}$ where $T(z)$ denotes a spanning tree of $J(z)$ for all $z \in Z$.*

Example 5.1 *Consider a graph $H^* = (R,D)$ given in Figure 5.1. Let $Z = \{a,c\} \subseteq R$ be a parameter set and (W,Z) be a soft set where the function $W : Z \to P(V)$ is defined by $W(z) = \{y \in R : d(z,y) \leqslant 1\}$ for all $z \in Z$. That is, $W(a) = \{a,b,f,e\}$ and $W(c) = \{b,c,d,e\}$. Let (Y,Z) be a soft set over D where the function $Y : Z \to P(E)$ is defined by $Y(z) = \{uv \in D : \{u,v\} \in W(z)\}$ for all $z \in Z$.*

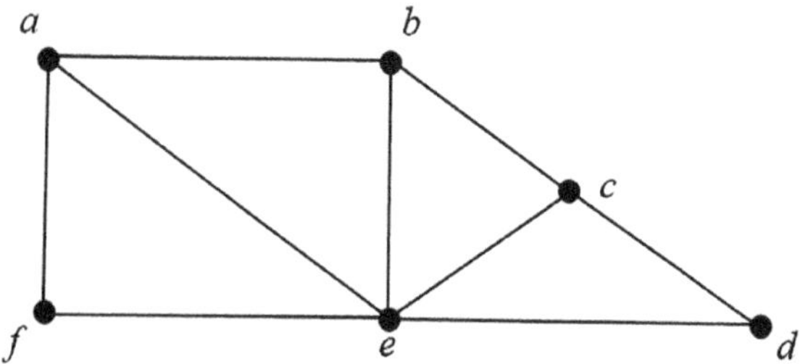

Figure 5.1 $H^* = (R,D)$.

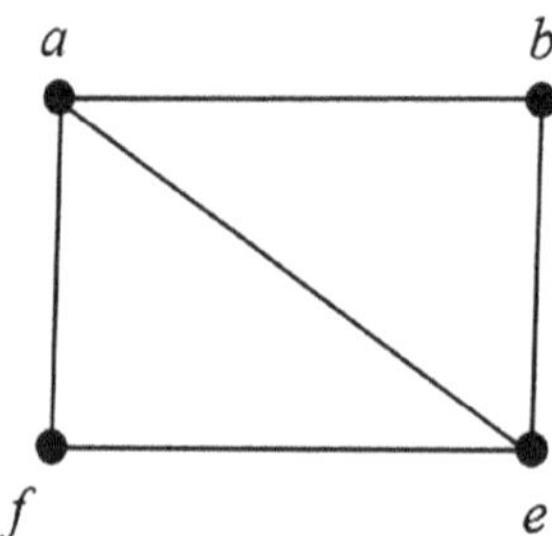

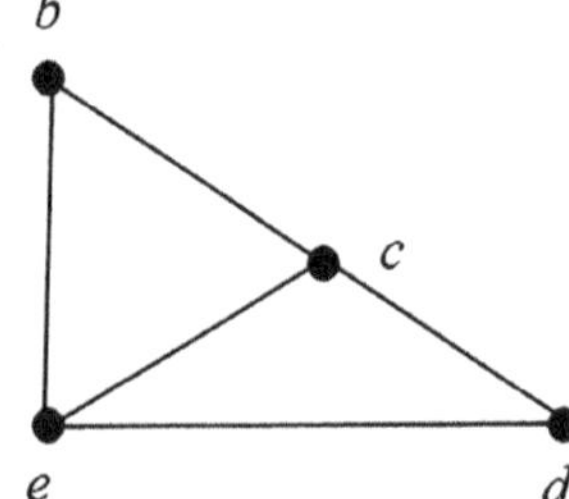

Figure 5.2 $H = \{J(a), J(c)\}$.

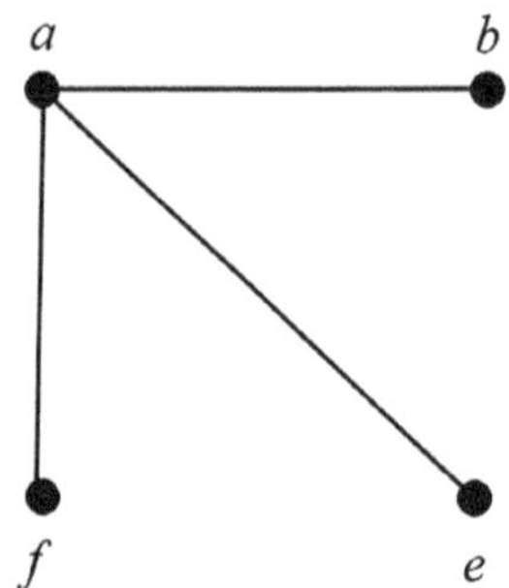

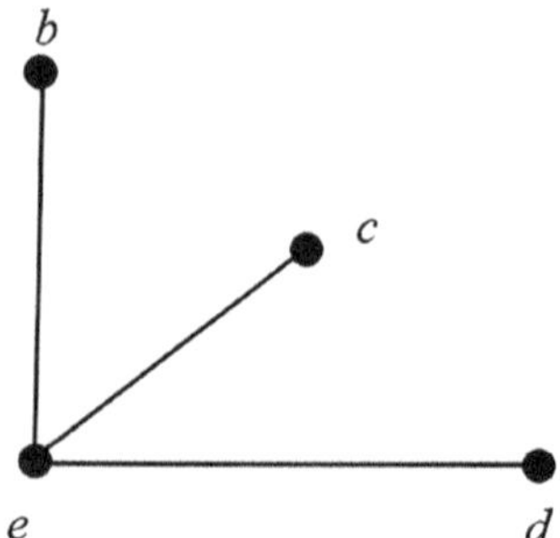

Figure 5.3 Spanning soft tree T_1 of $H = \{J(a), J(c)\}$.

That is, $Y(a) = \{ab, af, ae, be, ef\}$ and $Y(c) = \{bc, cd, be, ed, ce\}$. Thus, $J(a) = (W(a), Y(a))$ and $J(c) = (W(c), Y(c))$ are subgraphs of H^* as shown in Figure 5.2. Hence, $H = \{J(a), J(c)\}$ is a soft graph of H^*. A spanning soft tree T_1 of $H = \{J(a), J(c)\}$ is given in Figure 5.3. Another spanning soft tree T_2 of $H = \{J(a), J(c)\}$ is given in Figure 5.4. There are several other spanning trees for this soft graph H.

Theorem 5.1 *Let $H = (H^*, W, Y, Z)$ be a soft graph of the graph $H^* = (R, D)$ represented by $\{J(z) : z \in Z\}$. Then, the part $J(z)$ is connected for all $z \in Z$ if and only if H has a spanning soft tree.*

Proof *Assume that the part $J(z)$ of H is connected for all $z \in Z$. Consider any such connected part $J(z)$ for some $z \in Z$. If it contains no cycles, it will be a tree and also a spanning tree of $J(z)$. If $J(z)$ contains cycles, delete edges from the cycles without affecting the connectedness of $J(z)$ until no cycle remains in $J(z)$. Then, the new graph obtained like this will be a tree and is a spanning tree $T(z)$ of $J(z)$. Similarly, we can find the spanning trees $T(z)$ of the part $J(z)$ for all $z \in Z$. Then, $\{T(z) : z \in Z\}$ will be a spanning soft tree of the soft graph H. Conversely, assume that the soft*

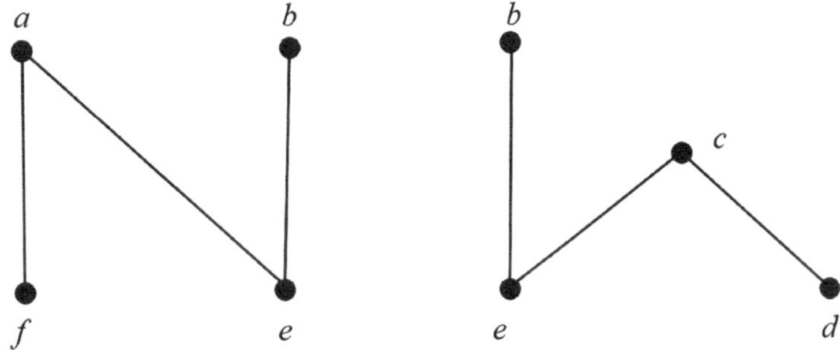

Figure 5.4   Spanning soft tree T_2 of $H = \{J(a), J(c)\}$.

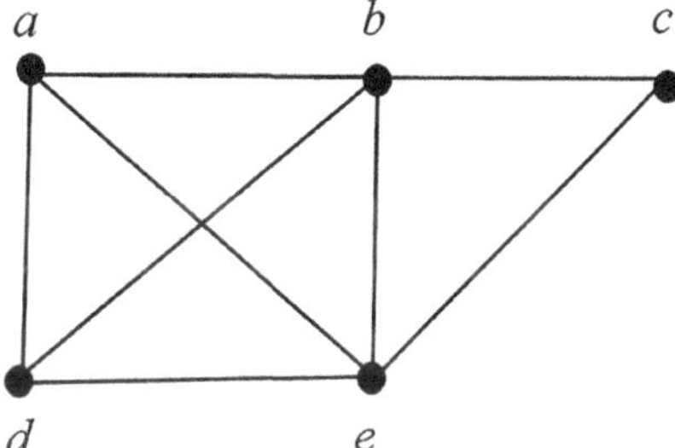

Figure 5.5 $H^* = (R, D)$.

graph $H = \{J(z) : z \in Z\}$ has a spanning soft tree $\{T(z) : z \in Z\}$. Then, $T(z)$ will be the spanning tree of the part $J(z)$ for all $z \in Z$. Definitely, the part $J(z)$ of H will be connected for all $z \in Z$.

Theorem 5.2 *A complete soft graph $H = (H^*, W, Y, Z)$ has $\Pi_{z \in Z}|W(z)|^{|W(z)|-2}$ different spanning soft trees.*

Proof *Let $H = (H^*, W, Y, Z)$ be a complete soft graph of a simple graph H^*. Then, the part $J(z)$ of H is complete for every $z \in Z$. Consider any such part $J(z)$ for some $z \in Z$. Since $J(z)$ is complete on $|W(z)|$ vertices by Cayley's theorem in graph theory, $J(z)$ has $|W(z)|^{|W(z)|-2}$ different spanning trees. The situations in all parts of H are the same. A spanning soft tree of H is given by $\{T(z) : z \in Z\}$ where $T(z)$ denotes a spanning tree of $J(z)$ for all $z \in Z$. So, if $Z = \{z_1, z_2 \cdots z_n\}$ (say), then the number of different spanning soft trees of H is given by $|W(z_1)|^{|W(z_1)|-2} \times |W(z_2)|^{|W(z_2)|-2} \times \cdots \times |W(z_n)|^{|W(z_n)|-2} = \Pi_{z \in Z}|W(z)|^{|W(z)|-2}$ since each part $J(z_i)$ has $|W(z_i)|^{|W(z_i)|-2}$ different possible drawings for spanning trees.*

Example 5.2 *Consider a graph $H^* = (R, D)$ given in Figure 5.5. Let $Z = \{a, c\} \subseteq R$ be a parameter set and (W, Z) be a soft set where the function $W : Z \to P(R)$ is defined by $W(z) = \{y \in R : wd(z, y) \leqslant 1\}$ for all $z \in Z$. That is, $W(a) = \{a, b, d, e\}$ and $W(c) = \{b, c, e\}$. Let (Y, Z) be a soft set over D where the function $K : Z \to P(E)$ is defined by $Y(z) = \{uv \in D : \{u, v\} \in W(z)\}$ for all $z \in Z$. That is, $Y(a) = \{ab, ad, ae, bd, be, de\}$ and $Y(c) = \{bc, be, ce\}$. Thus, $J(a) = (W(a), Y(a))$ and $J(c) = (W(c), Y(c))$ are subgraphs of H^* as shown in Figure 5.6. Hence, $H = \{J(a), J(c)\}$ is a soft graph of H^*. Here H has two parts $J(a)$ and $J(c)$ and both of them are complete. So, the soft graph H is a complete graph. Here $|W(a)| = 4$ and $|W(c)| = 3$. So, H has $4^{(4-2)} \times 3^{(3-2)} = 16 \times 3 = 48$ different spanning trees. But H has only $2 \times 1 = 2$ non-isomorphic spanning soft trees since $J(a)$ and $J(c)$ have only 2 and 1 non-isomorphic spanning trees, respectively. These two spanning soft trees T_1 and T_2 of H are given in Figures 5.7 and 5.8.*

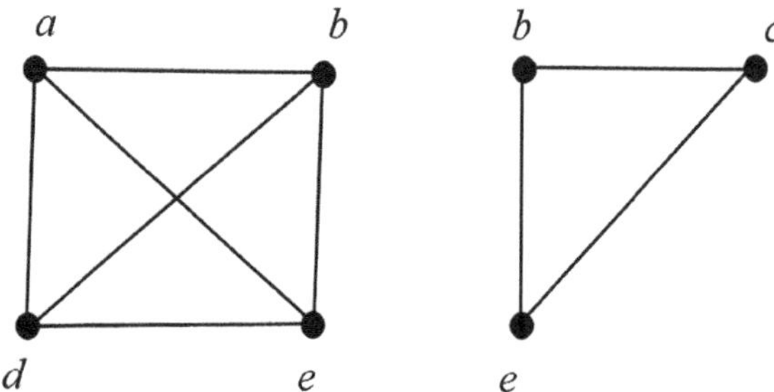

Figure 5.6 $H = \{J(a), J(c)\}$.

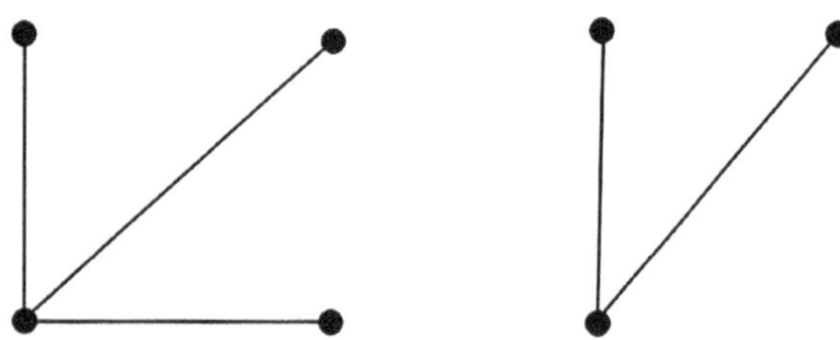

Figure 5.7   Spanning soft tree T_1 of $H = \{J(a), J(c)\}$.

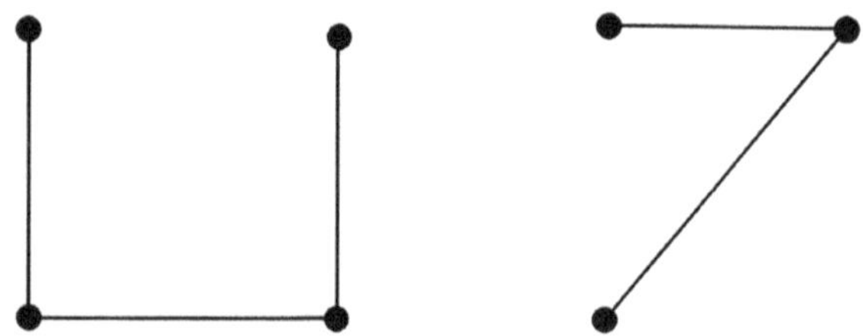

Figure 5.8   Spanning soft tree T_2 of $H = \{J(a), J(c)\}$.

Theorem 5.3 *If a soft graph H has a unique spanning soft tree, then H is a soft tree.*

Proof *Suppose that the soft graph H has a unique spanning soft tree $T = \{T(z) : z \in Z\}$. Assume that H is not a soft tree. Then definitely a part $J(z)$ of H for some $z \in Z$ has an edge $e = uv$ which is not contained in the corresponding spanning tree $T(z)$. Since $T(z)$ is a spanning subgraph of the corresponding part $J(z)$ and is a tree, there is a path P from u to v in $T(z)$. Then, $P + e$ is a cycle in the part $J(z)$. So, we can form a spanning tree $T'(z)$ which is different from $T(z)$ by adding the edge e and deleting an edge in the path P. Then, if we replace $T(z)$ for this particular $z \in Z$ by $T'(z)$, we get a spanning soft tree of H, which is different from T. This is a contradiction since T is unique. So, the soft graph H is a soft tree.*

Theorem 5.4 *Let $H^* = (R, D)$ be a simple graph and $H = (H^*, W, Y, Z)$ be a soft graph of H^* which is also represented by $\{J(z) : z \in Z\}$. Assume that the part $J(z)$ of H is $P_{|W(z)|}$ for every $z \in Z$ where $P_{|W(z)|}$ denotes a path on $|W(z)|$ vertices. Then, H has only one spanning soft tree.*

Proof *Since the part $J(z)$ of the soft graph H is a path on $|W(z)|$ vertices for every $z \in Z$, $J(z)$ is already a tree for every $z \in Z$. The deletion of any edge from $J(z)$ will disconnect the part $J(z)$. So, the spanning tree $T(z)$ of the part $J(z)$ is $J(z)$ itself. So, H has only one spanning soft tree which is $H = \{J(z) : z \in Z\}$ itself.*

Theorem 5.5 *Let $H^* = (R, D)$ be a simple graph and $H = (H^*, W, Y, Z)$ be a soft graph of H^* which is also represented by $\{J(z) : z \in Z\}$. If each part $J(z)$ is a connected subgraph of H^* and is not a tree, then H has at least $3^{|Z|}$ spanning soft trees.*

Proof *Let $J(z)$ be a part of the soft graph H for some $z \in Z$. Since $J(z)$ is a connected subgraph of H^* and is not a tree, then $J(z)$ contains a cycle, say $C(z)$. Since $J(z)$ is connected, it has definitely one spanning tree $T(z)$. As the tree is cycle-free, there is an edge in $C(z)$ that is not part of $T(z)$. Let that edge e connecting endpoints u and v. Here $T(z)$ includes both u and v. So, there exists a path $D(z)$ which connects u and v in $T(z)$ which does not include the edge e. We get another spanning tree $T_1(z)$ by deleting an edge e_1 on the path $D(z)$ and connecting u and v by e. We get another spanning tree $T_2(z)$ by deleting an edge e_1 on the path $D(z)$ and connecting u and v by e. We have the flexibility to select e_1 and e_2 as distinct edges because the path linking u and v in $T(z)$ includes a walk of length at least 2. If not, the graph will have multiple edges, and this is not possible since H^* is a simple graph. Hence, using this construction, $J(z)$ will have a minimum of three spanning trees. So, H has at least $3 \times 3 \times \dots |Z|$ times $= 3^{|Z|}$ spanning soft trees.*

Theorem 5.6 *Assume that the soft graph $H = (H^*, W, Y, Z)$ of the graph $H^* = (R, D)$ is a soft cycle. Then, H has $\Pi_{z \in Z}|W(z)|$ different spanning soft trees.*

Proof *Suppose that the soft graph $H = (H^*, W, Y, Z)$ is a soft cycle. Then, the part $J(z)$ of H is a cycle for every $z \in Z$. Consider such a part $J(z)$ for some $z \in Z$ which is a cycle on $|W(z)|$ vertices and $|W(z)|$ edges. A spanning tree of $J(z)$ will contain $|W(z)| - 1$ edges. So, we obtain $|W(z)|$ different spanning trees for the part $J(z)$, by deleting one edge each from the cycle. The situation is similar in all parts of H. So, H has totally $\Pi_{z \in Z}|W(z)|$ different spanning soft trees since each part has $|W(z)|$ different spanning trees.*

5.4 CONCLUSION

The application of soft set principles to graphs led to the emergence of soft graphs. Soft graph theory is gaining significant momentum in the field of discrete mathematics research due to its proficiency in managing parameterization. Soft graphs offer multiple ways to represent relationships within a graph using parameterization. In this study, we introduced the idea of spanning soft trees and explored various characteristics and properties associated with them. The term minimum spanning tree refers to a spanning tree with the smallest possible total weight. These trees are significant in a wide range of applications and algorithms, often finding use in electrical grids, water networks, and computer networks. The concept of spanning soft trees can be extended to minimum spanning soft trees and some of their applications.

REFERENCES

1. Abbood, M. A., Al-Swidi, A. A. J., & Omran, A. A. (2019). Study of Some Graph Types via Soft Graphs. *Journal of Engineering and Applied Sciences*, 14(8), 10375–10379.
2. Akram, M. & Nawaz, S. (2015). Operations on Soft Graphs. *Fuzzy Information and Engineering*, 7, 423–449.
3. Akram, M. & Nawaz, S. (2016). Certain Types of Soft Graphs. *UPB Scientific Bulletin, Series A*, 78(4), 67–82.
4. Akram, M. & Zafar, F. (2015). On Soft Trees, Buletinul Academiei de Stiinte a Republicii Moldova. *Mathematica*, 2(78), 82–95.
5. George, B., Jose, J., & Thumbakara, R. K. (2023). Tensor Products and Strong Products of Soft Graphs. *Discrete Mathematics, Algorithms and Applications*, 15(8), 1–28.
6. George, B., Jose, J., & Thumbakara, R. K. (2022). Co-normal Products and Modular Products of Soft Graphs. *Discrete Mathematics, Algorithms and Applications*, 16(2), 2350012.

7. Jose, J., George, B., & Thumbakara, R. K. (2023). Some Properties of Cartesian and Lexicographic Products of Soft Graphs. In: (eds.) Sharmistha Ghosh, Niranjanamurthy M, Krishanu Deyasi, Biswadip Basu Mallik, and Santanu Das, *Mathematics and Computer Science*, Volume 3. Scrivener Publishing: Beverly, MA.

8. Jose, J., George, B., & Thumbakara, R. K. (2023). *Eulerian Soft Graphs, Mathematics and Computer Science*, Volume 3. Scrivener Publishing: Beverly, MA.

9. Maji, P. K., Roy, A. R., & Biswas, R. (2001). Fuzzy Soft Sets. *The Journal of Fuzzy Math*, 9, 589–602.

10. Maji, P. K., Roy, A. R., & Biswas, R. (2002). An Application of Soft Sets in a Decision Making Problem. *Computers and Mathematics with Application*, 44, 1077–1083.

11. Molodtsov, D. (1999). Soft Set Theory-First Results. *Computers & Mathematics with Applications*, 37, 19–31.

12. Palani, K., Jones, T., & Maheswari, V. (2021). Soft Graphs of Certain Graphs. *Journal of Physics: Conference Series*, 1947(1), 1–7.

13. Sarala, N. & Manju, K. (2019). On Soft Bi-Partite Graph. *International Journal of Basic and Applied Research*, 9(7), 249–256.

14. Sarala, N. & Manju, K. (2019). Some Types of Domination in Soft Graph. *International Journal of Research and Analytical Reviews*, 6(2), 968–973.

15. Sarala, N. & Manju, K. (2019). Detour Domination in Soft Graph. *International Journal of Scientific & Technology Research*, 8(9), 774–777.

16. Thenge, J. D., Reddy, B. S., & Jain, R. S. (2019). Contribution to Soft Graph and Soft Tree. *New Mathematics and Natural Computation*, 15(1), 129–143.

17. Thenge, J. D., Reddy, B. S., & Jain, R. S. (2020). Connected Soft Graph. *New Mathematics and Natural Computation*, 16(2), 305–318.

18. Thenge, J. D., Reddy, B. S., & Jain, R. S. (2020). Adjacency and Incidence Matrix of a Soft Graph. *Communications in Mathematics and Applications*, 11(1), 23–30.

19. Thenge, J. D., Reddy, B. S., & Jain, R. S. (2021). Application of a Soft Graph in Decision Making Using Adjacency Matrix of a Soft Graph. *Ganita*, 71(2), 181–188.

20. Thumbakara, R. K. & George, B. (2014). Soft Graphs. *General Mathematics Notes*, 21(2), 75–86.

21. Thumbakara, R. K., George, B., & Jose, J. (2022). Subdivision Graph, Power and Line Graph of a Soft Graphs. *Communications in Mathematics and Applications*, 13(1), 75–85.

22. Thumbakara, R. K., Jose, J., & George, B. (2022). Hamiltonian Soft Graphs. *Ganita*, 72(1), 145–151.

23. Thumbakara, R. K., George, B., & Jose, J. (2023). On Soft Graph Isomorphisms. *New Mathematics and Natural Computation*, 1–17.

24. Venkatraman, S. & Helen, R. (2019). On Domination in Soft Graph of Some Special Graphs. *Malaya Journal of Matematik*, S(1), 527–531.

25. Venkatraman, S., Helen, R., & Natarajan, C. (2019). A New Approach to Domination in Soft Graphs. *International Journal of Applied Engineering Research*, 14(4), 35–45.

6 A Review on the Transshipment Problems in Optimization Theory

Karampreet Kaur and Sandeep Singh

The transshipment problem is a certain type of transportation problem. It is a linear programming problem in which shipping paths cover intermediate points and find the minimum shipping cost. However, many real-life problems involve uncertainty, and to address these challenges, uncertainty needs to be incorporated into transshipment problems. The transshipment problem under uncertainty is a transshipment problem in which parameters such as transshipment cost, supply, and demand quantities are not crisp but some are fuzzy, or maybe all are fuzzy. The main objective of this chapter is to focus on the work done by various researchers on the transshipment problems in order to expand the different variants of this problem in the future.

6.1 INTRODUCTION

The transshipment problem is a variant of the classical transportation problem and is used to determine the distribution of goods or resources from multiple sources to multiple destinations, allowing intermediate points (transshipment points) where goods can be temporarily stored or transferred. It has many applications such as supply chain management, inventory management, and network design. In the current intense commercial competition, organizations face mounting pressure to discover more effective methods for producing and delivering products and services to meet customer demands. Determining the optimal timing and cost-effective means of delivering products to customers in their desired quantities has become increasingly complex. Transshipment models offer a robust framework to address this challenge and ensure the efficient movement and timely availability of raw materials and finished goods. But, many real-life problems involve uncertainty, and to address these challenges, uncertainty needs to be merged into transshipment problems. The transshipment problem under uncertainty extends the traditional transshipment problem by considering uncertainty in input data, where parameters like costs, demands, and supplies are represented as fuzzy numbers. The transshipment problem under uncertainty is particularly valuable when dealing with real-world situations where data are not precise but is represented with degrees of uncertainty, helping decision-makers make more informed and flexible choices. This chapter is organized as follows: Section 6.2 presents types of transshipment problems without any uncertainty such as classical transshipment problems, fixed charge transshipment problems, inventory transshipment problems, and multi-objective transshipment problems, with their mathematical models. Section 6.3 covers the survey on recent developments of transshipment problems under uncertainty and the last section presents the conclusions.

6.2 TYPES OF TRANSSHIPMENT PROBLEM

The transshipment problem is a class of optimization problems in which goods or commodities are moved from an origin to a destination through one or more intermediate nodes or transshipment points. There are several variations of transshipment problems, each with its specific characteristics. The types of transshipment problems include the following.

DOI: 10.1201/9781003407386-6

6.2.1 CLASSICAL TRANSSHIPMENT PROBLEM

It is the most basic form of the transshipment problem, where goods can be transferred between intermediate nodes, also known as transshipment nodes, before reaching their final destination. The objective is to minimize transportation costs while meeting supply and demand requirements. The mathematical formulation of the classical transshipment problem is as follows:
Objective function:

$$\text{Minimize } Z = \sum_{i \in I} \sum_{j \in J} c_{ij} x_{ij}$$

Supply constraints:

$$\sum_{j \in J} x_{ij} + \sum_{k \in K} y_{ik} = s_i \quad \text{for all } i \in I$$

Demand constraints:

$$\sum_{i \in I} x_{ij} + \sum_{k \in K} z_{kj} = d_j \quad \text{for all } j \in J$$

Non-negative constraints:

$$x_{ij} \geq 0 \quad \text{for all } i \in I, j \in J$$
$$y_{ik} \geq 0 \quad \text{for all } i \in I, k \in K$$
$$z_{kj} \geq 0 \quad \text{for all } k \in K, j \in J$$

where
c_{ij} Cost of shipping one unit from supply point i to demand point j,
x_{ij} The quantity of goods shipped from the supply point i to demand point j,
y_{ik} The quantity of goods transshipped from the supply point i to transshipment point k,
z_{kj} The quantity of goods shipped from transshipment point k to demand point j,
d_j Demand at demand point j,
s_i Supply at supply point i.

6.2.2 FIXED CHARGE TRANSSHIPMENT PROBLEM

There are fixed costs associated with using transshipment nodes. The objective is to minimize the total transportation and fixed costs while satisfying supply and demand constraints.

The mathematical formulation of the fixed charge transshipment problem is as follows:
Objective function:

$$\min \sum_{i \in N} \sum_{j \in N} C_{ij} x_{ij} + \sum_{k \in N} F_k y_k$$

Flow conservation constraints:

$$\sum_{j \in N} x_{ij} - \sum_{j \in N} x_{ji} = D_i \quad \text{for all } i \in N$$

Capacity constraints:

$$x_{ij} \leq U_{ij} \quad \text{for all } (i, j) \in A$$

Transshipment node constraints:

$$x_{ik} \leq U_{ik} y_k \quad \text{for all } (i,k) \in N$$

Non-negative constraints:

$$x_{ij} \geq 0 \quad \text{for all } (i,j) \in A$$
$$y_k \in \{0,1\} \quad \text{for all } k \in N$$

where
C_{ij} Cost of shipping one unit of commodity from node i to node j (arc-based transportation cost),
x_{ij} Flow of commodity from node i to node j through arc (i,j),
y_k Binary variable indicating whether transshipment node k is open or closed,
F_k Fixed cost associated with transshipment node k,
D_i Demand at node i,
N Set of nodes,
A Set of arcs connecting the nodes.

6.2.3 TIME-CONSTRAINED TRANSSHIPMENT PROBLEM

The time-constrained transshipment problem introduces time constraints on the transportation process. It considers the time it takes to transport goods from one node to another, which can vary for different routes.

The mathematical formulation of the time-constrained transshipment problem is as follows:
Objective function:

$$\min \sum_{i \in N} \sum_{j \in N} C_{ij} x_{ij}$$

Flow conservation constraints:

$$\sum_{j \in N} x_{ij} - \sum_{j \in N} x_{ji} = 0 \quad \text{for all } i \in N$$

Capacity constraints:

$$x_{ij} \leq U_{ij} \quad \text{for all } (i,j) \in A$$

Time constraints:

$$L_i \leq t_i \leq U_i \quad \text{for all } i \in N$$

Arrival time calculation:

$$t_i = t_j + \frac{x_{ij}}{U_{ij}} \quad \text{for all } (i,j) \in A$$

Time constraint satisfaction:

$$t_i \leq T_i \quad \text{for all } i \in N$$

Non-negative constraints:

$$x_{ij} \geq 0 \quad \text{for all } (i,j) \in A$$

where

C_{ij} Cost of shipping one unit of commodity from node i to node j (arc-based transportation cost),

x_{ij} Flow of commodity from node i to node j through arc (i,j),

U_{ij} Capacity of arc (i,j),

T_i Time constraint at node i,

t_i Arrival time at node i,

L_i Time window lower bound at node i,

U_i Time window upper bound at node i,

N Set of nodes,

A Set of arcs connecting the nodes.

6.2.4 HUB LOCATION TRANSSHIPMENT PROBLEM

In this problem, there is a network of hubs, and goods can be transshipped at these hubs. The objective is to determine the locations of hubs, as well as the routing of goods through the hubs, to minimize overall transportation costs.

The mathematical formulation of the hub location transshipment problem is as follows:
Objective function:

$$\min \sum_{j=1}^{n} f_j x_j + \sum_{i=1}^{m} \sum_{j=1}^{n} c_{ij} y_{ij}$$

Constraints:

$$\sum_{j=1}^{n} y_{ij} = 1 \quad \text{for all } i \in I$$

$$\sum_{j=1}^{n} x_j \leq B$$

$$\sum_{i=1}^{m} y_{ij} \leq M x_j \quad \text{for all } j \in J$$

Non-negative constraints:

$$x_j \in \{0,1\} \quad \text{for all } j \in J$$
$$y_{ij} \in \{0,1\} \quad \text{for all } i \in I, j \in J$$

where

c_{ij} Transportation cost per unit between customer location i and hub location j (non-negative real number),

x_j Takes the value 1 if a hub is opened location j, 0 otherwise (binary decision variable),

y_{ij} Takes the value 1 if customer i is allocated to hub j, 0 otherwise (binary decision variable),

f_j Fixed cost associated with selecting hub location j,

d_{ij} The demand at the customer location i (positive integer),

I Set of customer locations (indexes $i = 1, \ldots, m$),

J Set of potential hub locations (indexes $j = 1, \ldots, n$).

6.2.5 INVENTORY TRANSSHIPMENT PROBLEM

The inventory transshipment problem combines the transshipment problem with inventory management. It involves determining the optimal levels of inventory at transshipment points while satisfying supply and demand requirements.

The mathematical formulation of the inventory transshipment problem is as follows:
Objective function:

$$\min \sum_{i \in I} \sum_{j \in J} X_{ij} C_{ij}$$

Demand satisfaction:

$$\sum_{i \in I} X_{ij} = d_j \quad \text{for all } j \in J$$

Inventory balance:

$$\sum_{i \in I} X_{ij} - \sum_{j \in J} X_{ji} = I_i \quad \text{for all } i \in I$$

Inventory bounds:

$$L_i \leq I_i \leq U_i \quad \text{for all } i \in N$$

Non-negative constraints:

$$X_{it} \geq 0 \quad \text{for all } i \in I, j \in J$$

where
C_{ij} Cost of transshipping one unit of inventory from source i to destination j,
X_{ij} The amount of inventory transshipped from source i to destination j,
I_i Initial inventory level at source i
U_i The upper bound or maximum capacity of inventory that can be stored at supply/transshipment node i,
L_i The lower bound or minimum level of inventory that can be stored at supply/transshipment node i,
d_j Demand at destination j,
I Set of source locations
J Set of destination locations.

6.2.6 MULTI-OBJECTIVE TRANSSHIPMENT PROBLEM

In this case, multiple conflicting objectives are considered simultaneously, such as minimizing costs and minimizing travel time. The problem involves finding trade-offs between these objectives.

The mathematical formulation of the multi-objective transshipment problem is as follows:
Objective function 1:

$$\text{Minimize} : f_1(I, x)$$

Objective function 2:

$$\text{Minimize} : f_2(I, x)$$

$$\cdots$$

Objective function 3:

$$\text{Minimize}: f_k(I, x)$$

Demand satisfaction constraints:

$$\sum_{i \in I} x_{ij} = d_j \quad \text{for all } j \in J$$

Inventory balance constraints:

$$\sum_{j \in J} x_{ij} - \sum_{j \in J} x_{ji} = I_i \quad \text{for all } i \in I$$

Inventory bounds:

$$L_i \le I_i \le U_i \quad \text{for all } i \in I$$

Non-negative constraints:

$$x_{ij} \ge 0 \quad \text{for all } i \in I, j \in J$$
$$I_i \ge 0 \quad \text{for all } i \in I$$

where
c_{ij} Cost of transshipping one unit of inventory from source i to destination j,
x_{ij} The amount of inventory transshipped from node i to node j,
I_i Initial inventory level at source i
U_i The upper bound or maximum capacity of inventory that can be stored at supply/transshipment node i,
L_i The lower bound or minimum level of inventory that can be stored at supply/transshipment node i,
d_j Demand at destination j,
I Set of source locations
J Set of destination locations.

6.2.7 TRANSSHIPMENT PROBLEMS WITHOUT ANY UNCERTAINTY

This section provides a survey of the existing research on transshipment problems without any uncertainty. Numerous authors have focused their efforts on studying these problems and have suggested methodologies for resolving them.

In 1956, Orden [46] introduced a mathematical approach that can be extended, along with pairwise configurations for finding the optimal interconnected routes across a collection of points. In 1964, King and Logan [30] presented an issue that revolves around determining the placement and dimensions of cattle slaughtering facilities in California, considering the positioning and volume of animals to be slaughtered, as well as the demand for the final products. An iterative approach has been employed to integrate the advantages of scale in production, alongside the expenses associated with transferring goods, to achieve the most cost-effective resolution. In 1965, Hurt and Tramel [19] presented an alternative problem formulation that has been presented using the transportation model framework. These formulations enable the resolution of problems similar to those examined by King and Logan, all without requiring the subtraction of artificial variables.

In 1977, Hoadley and Heyman [20] introduced a single-period inventory model comprising two echelons, consisting of a primary warehouse in the first echelon and secondary warehouses in the second echelon. The study revealed that the objective function exhibits one of four distinct structures, determined by the relative scales of the shipping costs, and it has been established that all four variations of the objective function demonstrate convexity. In 1987, Lee [38] presented a continuous-review, multi-echelon model designed for repairable items, where emergency lateral transshipments between identical bases are permitted. The study derives approximations and employs them to ascertain the most efficient stocking levels within this system, offering a step-by-step procedure to calculate these levels. In 1990, Hoppe [21] introduced the first polynomial-time technique for addressing the quickest transshipment problem while making significant contributions to the attainment of integral optimal outcomes. In 1998, Needham et al. [43] developed an approach for logistics managers to reduce inventories while maintaining customer service levels through emergency transshipments. The paper analyzes the financial impact of these decisions.

In 1999, Herer and Rashit [22] focused on the problem of inventory control in a two-location inventory system, with transshipments being employed as substitute supply mechanisms once demand events occur, and in 2001, Herer [23] discussed the application of transshipment strategies in a dynamic and high-priority demand setting within a finite planning timeframe. This research represents the first-ever exploration of transshipment techniques in a dynamic or critical context. In the same year, Fleischer [12] presented an algorithm that finds the quickest transshipment with a polynomial number of maximum flow computations and a faster algorithm that also uses minimum cost flow computations. When there was only one sink, we show how the algorithm can be sped up to return a solution using $O(k)$ maximum flow computations, where k is the number of sources. In 2002, Herer et al. [24] developed a tactical strategy for achieving leagility without resorting to postponement in the context of transshipments, a prevalent practice in multi-location inventory systems for transferring stock within the supply chain. In 2004, Sarker and Namatame [52] presented the three evolutionary algorithms related to binary and real coding that has been outlined with the aim of addressing a transshipment problem. Additionally, it introduces a customary optimization software package for comparing the solution's quality.

In 2005, Hu et al. [25] presented a multiple-retailer distribution system for emergency transshipments. To find the effect of transshipments on ordering policies, a solution procedure based on the dynamic programming technique has been developed. In 2007, Lee et al. [39] investigated a novel lateral transshipment strategy, referred to as the Service Level Adjustment (SLA) policy, which has been introduced. The proposed policy offered reduced costs compared to previous transshipment strategies. In the same year, Keskin [31] aimed to depict the distribution of 50 cl Pepsi soft drinks as a transshipment challenge. The primary objective of this study was to ascertain the most economical means of conveying 10,000 crates of the commodity from the Benin manufacturing facility (origin) via intermediary depots (transshipment hubs) to meet the demand for the product in the Sapele-Warri area (destinations). Data collected were analyzed using TORA Windows Version 2.00 software. In 2006, container shipping marked its 50th anniversary as a groundbreaking innovation that had an enormous influence on the worldwide landscape of production and distribution. The globalization of production was facilitated by a more effective utilization of comparative advantages, while distribution networks were able to engage in more streamlined interactions. The mounting challenges within container logistics within the context of global supply chains were analyzed by Notteboom and Rodrigue [44] in 2008. In the same year, Agarwal [2] introduced a comprehensive model that combines ship-scheduling and cargo-routing problems, addressing them concurrently. To solve the problem, the authors devised a greedy heuristic, a column generation-driven algorithm, and a two-phase Benders decomposition-centered algorithm.

In 2010, Bris [8] introduced a transshipment problem within the broader context of the MANY-TO-MANY logistics problem, specifically focusing on scenarios with a single transshipment point. In 2011, Reinhardt and Pisinger [51] represented the challenges associated with network design and fleet assignment have been integrated into a mixed integer linear programming (MILP) model with the objective of minimizing the total expenses. As far as our understanding goes, this is the inaugural instance where a precise solution approach for this issue takes into account the costs associated with transshipment. It was the first time an exact solution method to the problem considers transshipment cost. The problem has been successfully addressed utilizing a branch-and-cut method, featuring the utilization of clover and transshipment inequalities.

In 2013, Khurana and Arora [32] introduced a technique for addressing transshipment problems. The problem was solved by transforming the original problem into a comparable transportation problem in order to determine the most efficient solution. The paper discussed both balanced and unbalanced transshipment problems. In 2014, Noham and Tzur [45] focused on supply chain systems that incorporate lateral transshipments. Specifically, in a system featuring two retailers confronting stochastic demand, we move away from the presumption of negligible fixed transshipment costs. This expansion of existing findings pertains to the single-item scenario and introduces a novel model encompassing multiple items. A historical analysis was conducted on research conducted over the past three decades, where operations research techniques have been employed to address challenges in containership routing and scheduling across strategic, tactical, and operational planning domains. They focused on the categorization and summary of these issues, with a particular emphasis on the structure of mathematical models, underlying assumptions, and the methodologies designed for problem-solving by Meng et al. [40] in 2014. In the same year, Alkhulaifi et al. [3] introduced a technique for addressing a particular type of bi-criteria multistage transportation problem with transshipment. This method can also be applied to bi-criteria single-stage transportation problems. In 2015, Khurana [34] presented defective and enhanced flow in a transshipment problem, taking inspiration from the problem itself and discussed capacitated transshipment problems that incorporate constraints on the total supply at the source and the total demand at the destination.

In 2016, Ashioba and Nwachukwu [4] introduced the enhanced neural network model to discover the optimal solution for a transshipment problem. This model is capable of accommodating both normalized and un-normalized data sets as input. In 2017, Kumari et al. [35] developed a novel and straightforward strategy, the Max-Min method was presented to solve transshipment problems that include mixed constraints. This method can be employed for all types of transshipment problems. In 2019, Wichapa and Khokhajaikiat [63] discussed a multi-criteria transshipment problem associated with infectious waste management (IWM). The primary goal is to identify a new transshipment network for IWM. The study includes a real-world case scenario involving forty hospitals and three prospective disposal municipalities in Northeastern Thailand. The main objective of the work done by Agadaga and Akpan [6] in 2019 was to achieve the most cost-effective means of transporting 10,000 crates of the item from the Benin manufacturing plant through supply centers to the destinations where the product is needed.

In 2021, Crainic [10] explored the synchronized location transshipment challenge, where the essential determinant is the careful selection of facilities to preserve flow synchronization. A multi-objective transshipment problem for multiple items has been contained. In the study, a technique was presented to reduce travel time while simultaneously maximizing profit studied by Aishwarya [7] in 2021. In 2022, Sultan and Alsaber [56] constructed a multi-objective framework using a priori preference techniques to attain a satisfactory outcome across all objectives in a transshipment challenge that includes different vehicle categories and routing scenarios. Explored a comprehensive issue involving transshipment for cross-filling and the reconciliation of missing demand data by Vu and Ko [62] in 2022. The paper addressed the logistics of conducting transshipment operations when dealing with gaps in demand data. A unique integrated problem has been introduced to combine the processes of imputing missing demand data with strategic transshipment decision-making. In the same year, Zhou and Wang [65] discussed about small businesses with many stocking locations face high ordering costs and imbalanced inventories. In response to these difficulties, an inventory management system was developed for multiple retailers. The research presented an analytical framework for assessing the associated costs, established constraints to identify the ideal ordering interval, order-up-to-level, and timing for transshipment and determined transshipment quantities based on marginal costs.

The dynamic pickup and delivery problem, which incorporates elements like random orders, selection of transshipment locations, and adherence to last-in-first-out (LIFO) constraints, adds complexity to distribution optimization and routing planning. Consequently, the paper developed a multi-objective mathematical model designed to address dynamic pickup and delivery problems featuring transshipments and LIFO constraints (DPDPTL). The model's primary objectives are to minimize driving distance and maximize order satisfaction, thereby tackling the challenges essential to this problem by Xu and Wei [64] in 2023. In the same year, Su et al. [57] developed a novel variation of the vehicle routing problem within the context of last-mile delivery, known as the Crowd-Shipping Problem with Time Windows, Transshipment Nodes, and Delivery Options (CSPTW-TN-DO). This problem involves the utilization of two distinct types of fleets, specifically dedicated vehicles and occasional drivers, to cater to three different customer categories. In the same year, Skutella [59] showed that the least cost quickest flow problem falls into the category of NP-hard problems. The paper clarifies the transition of the quickest minimum cost transshipment problem into a more easily solvable quickest transshipment problem.

6.3 TRANSSHIPMENT PROBLEMS UNDER UNCERTAINTY

Transshipment problems under uncertainty are optimization problems that involve the allocation of goods or resources between different supply and demand nodes in a network while considering various sources of uncertainty. These problems are encountered in supply chain and logistics management, and they often require decision-making that can adapt to the effects of uncertainty. Several types of uncertainty can be considered in transshipment problems, including demand uncertainty, cost uncertainty, and capacity uncertainty.

Mathematical formulation of transshipment problems under uncertainty: In transshipment problems, the discrimination between a source and destination is dropped so that a transportation problem with m sources and n destinations gives an increment to a transshipment problem with $m + n$ sources and $m + n$ destinations. The basic feasible solution to such a problem will concern $[(m+n)+(m+n)-1]$ or $2m+2n-1$ basic variables, and if we discard the variables emerging in the $(m+n)$ diagonal cells, we are left with $m+n-1$ basic variable. Thus, the transshipment problems under uncertainty are written as:

$$\text{Minimize } \tilde{z} \approx \sum_{i=1}^{m+n} \sum_{i=1, j \neq i}^{m+n} \tilde{c}_{ij} \tilde{x}_{ij} \quad \text{subject to}$$

$$\sum_{j=1, j \neq i}^{m+n} \tilde{x}_{ij} - \sum_{j=1, j \neq i}^{m+n} \tilde{x}_{ji} \approx \tilde{a}_i, i = 1, 2, \ldots, m$$

$$\sum_{i=1, i \neq j}^{m+n} \tilde{x}_{ij} - \sum_{i=1, i \neq j}^{m+n} \tilde{x}_{ji} \approx \tilde{b}_j, j = m+1, m+2, \ldots, m+n$$

$$\tilde{x}_{ij} \succcurlyeq \tilde{0}, i, j = 1, 2, \ldots, m+n, j \neq i$$

where
$\tilde{c}_{ij}$ the fuzzy transshipment cost of the product from the ith source to jth destination,
$\tilde{x}_{ij}$ the fuzzy quantity of the product that is transported from the ith source to jth destination,
$\tilde{b}_j$ the fuzzy demand quantity at destination j,
$\tilde{a}_i$ the fuzzy supply quantity by origin i,
m the number of sources,
n the number of destinations.

There are various forms of transshipment problems under uncertainty like interval transshipment problems, multi-objective transshipment problems, dynamic transshipment problems, bi-level transshipment problems, and robust transshipment problems. In the mathematical formulation of the types of transshipment problems under uncertainty, the variables such as supply and demand quantities, cost coefficients, and decision variables are uncertain, and the sign '$\sim$' (tilde) denotes the uncertainty. These types of transshipment problems address the complexities of real-world scenarios where decision-makers must contend with imperfect or uncertain information. Solving these problems typically involves advanced optimization techniques, simulation, and risk analysis to arrive at effective solutions.

In 1990, Perincherry [47] developed a fuzzy linear programming framework aimed at addressing the transshipment challenge. The primary aim was to identify a shipping schedule capable of efficiently conforming to cost limitations. In 2009, Hmiden et al. [26] developed a hybrid algorithm that combines fuzzy simulation and genetic algorithms to establish the best transshipment strategy. This algorithm integrates both pessimistic and optimistic decision-maker perspectives to identify the exact timing and quantity for transshipments. In 2010, Said et al. [53] presented the transshipment approach for a two-stage transshipment model involving uncertain or fuzzy demands. In 2011, Javaid and Gupta [29] discussed a stochastic capacitated transshipment problem in the presence of uncertain demand for consignment transportation. It formulated an optimal placement for the transshipment facilities aimed at minimizing the total cost. In the same year, Gani [13] solved the fuzzy transshipment problem to minimize the fuzzy cost using Fuzzy Vogel's approximation method. A novel approach is introduced for determining the optimal solution of fuzzy transportation problems involving transshipment, which encompasses the following scenarios: (i) Transfer from one source to another source, (ii) relocation between different destinations, and (iii) movement from a destination to any source, by Kumar et al. [33] in 2011.

In 2012, Baldi [9] introduced the capacitated transshipment location problems under uncertainty (CTLP), which involve a two-stage stochastic program with recourse, have been transformed into an equivalent deterministic non-linear capacitated transshipment location problem (CTLP). This specific problem falls under the category of entropy-maximizing models, which belong to a broad range of mathematical formulations. The performance of CTLP is highly satisfactory. In the same year, Rajendran and Pandian [50] developed a novel approach, referred to as the "splitting method," which was introduced to find an optimal solution for a transshipment problem characterized by real interval parameters. Furthermore, the splitting method has been expanded to address fully fuzzy transshipment problems.

In 2014, Sirajudeen and Gani [54] presented an intuitionistic fuzzy transshipment problem that was transformed into a transportation problem and then solved by Fuzzy Vogel's approximation method. In the same year, Gani et al. [14] dealt with the large-scale transshipment problem in a fuzzy environment. The performance of an improved version of Vogel's approximation method over Vogel's approximation method has been discussed. In 2015, Gayathri et al. [15] examined the Max-Min method as a means to derive an initial feasible solution for large-scale transshipment problems that involve triangular fuzzy numbers. To convert these triangular fuzzy numbers into precise values, the graded mean integration method was employed. In 2016, Prabha and Vimala [48] introduced a magnitude approach for transshipment problems involving trapezoidal fuzzy numbers. They conducted a comparative analysis of solutions generated through Vogel's approximation method (VAM), least cost method (LCM), and north-west corner method (NWCM) to find the best optimal solution. Among these techniques, VAM yielded the most optimum solution. In the same year, Gayathri and Subramanian [16] examined a transshipment problem in which all parameters were represented by trapezoidal fuzzy numbers. The authors proposed an alternative methodology that differs from existing techniques and offered a unique perspective on handling transshipment problems under conditions of uncertainty.

In 2017, Mohanapriya and Jeyanthi [41] used Vogel's approximation method (VAM) to find an efficient initial solution for the large-scale transshipment problem. During the same year, Islam and Ray [28] solved a multi-objective triangular transshipment problem under uncertainty.

In 2018, Alp et al. [5] focused on the multi-objective transshipment problem, which was formulated using the fuzzy goal programming approach developed by Human and Chen, and it was resolved using the WinQSB program. In the same year, Hamoud [27] introduced a concept of generalized hexagonal fuzzy numbers for addressing fuzzy transshipment problems. The primary objective was to determine the optimal solution for solving transshipment challenges. We addressed fuzzy multi-objective transportation problems and achieved a fuzzy optimal compromise solution through the application of a fuzzy genetic algorithm by Thiagarajan and Kandasamy [60] in 2018. In 2019, Vidhya and Ganesan [61] developed a Max-Min technique for attaining the initial feasible solution for the transshipment problem with trapezoidal fuzzy numbers. They evaluated the optimality of the problem using a fuzzy iteration of the modified distribution method.

In 2020, Giusti et al. [17] proposed a two-step stochastic programming formulation with an alternative, focusing on a set of instances. It utilizes well-established economic indicators to derive managerial conclusions concerning the importance of managing uncertainty within the problem context. In the same year, Kumar et al. [36] introduced a procedure based on the value and ambiguity-driven ranking methodology. This method directly achieved a fuzzy optimal result, eliminating the need for an initial feasible solution for the transshipment problem involving trapezoidal intuitionistic fuzzy numbers. Mohanaselvi and Prabakaran [42] in 2020 explored the domain of an uncertain transshipment problem, where the flow can vary with impairments or enhancements, and detailed a procedure for finding the optimal solution. During the same year, Sujatha [55] involved in the transformation of a transshipment problem featuring fuzzy parameters into a standard fuzzy transportation model. Hence, it is solved using the fuzzy one-point method in k-stages. In 2021, Ellaimony et al. [11] introduced a mathematical modeling strategy for addressing the transshipment problem with imprecise cost coefficients. They defined the issue as a quadratic mixed-integer linear programming problem. To address this challenge, the authors made use of the α-level set of fuzzy numbers as a methodological technique. In the same year, Garg et al. [18] presented a technique for solving the fuzzy fractional two-stage transshipment problem by utilizing a decomposition-based algorithm, which is used to address two distinct sub-problems to ascertain lower and upper bounds for the primary problem.

In 2022, Kumar et al. [37] presented a neutrosophic transshipment problem that has been introduced, featuring the expression of all parameters using trapezoidal neutrosophic numbers for the first time. The proposed technique makes use of the possibility of the mean ranking function in handling the neutrosophic transshipment problem. In 2023, Akilbasha et al. [1] proposed a study to solve fully fuzzy interval integer transshipment (FIIT) problems. The back-order sequence method has been formulated to find the optimal solution to the (FIIT) problem. Praveena and Vinotha [49] introduced a novel approach to address the multi-objective transshipment problem within a neutrosophic framework in the same year. The proposed method is straightforward and easy to grasp, offering improved solutions in both precise and uncertain scenarios. All the requirements, provisions, and shipping expenses are expressed as single-valued trapezoidal neutrosophic numbers. In 2023, Sarkar et al. [58] employed C++ code to implement the Fourier elimination method, which involves systematically eliminating variables one by one while retaining the objective function. This approach incorporated both impaired and enhanced flow conditions within the proposed model. In this model, the supply is treated as a fuzzy variable, allowing for flexibility, while the demand is considered a random variable, subject to variations based on physical phenomena.

6.4 CONCLUSION

This chapter provides a survey of recent developments in transshipment problems such as classical transshipment problems, fixed charge transshipment problems, time-constrained transshipment problems, inventory transshipment problems, and multi-objective transshipment problems with their mathematical models.

REFERENCES

1. Akilbasha, A., Natarajan, G., & Pandian, P. (1965). Optimising fully fuzzy interval integer transshipment problems. *International Journal of Operational Research, Sustainability*, 46(1), 1–19.
2. Agarwal, R., & Ergun, O. (2008). Ship scheduling and network design for Cargo routing in liner shipping. *Transportation Science*, 42(2), 127–261.
3. Alkhulaifi, K., AlRajhi, J., Elsayed R., Ellaimony, E. M., AlArdhi, M., & Abdelwali, H. A. (2014). An algorithm for solving bi-criteria large scale transshipment problems. *Global Journal of Researches in Engineering: B Automotive Engineering*, 14(4), 1–2.
4. Ashioba, N. C., & Nwachukwu, E. O. (2016). Transshipment problem using modified neural network model. *International Journal of Engineering Research and Technology IJERT*, 5, 83–87.
5. Alp, S., & Ozkan, T. K. (2018). Modeling of multi-objective transshipment problem with Fuzzy goal programming. *International Journal of Transportation*, 6(2), 9–20.
6. Agadaga, G. O., & Akpan, N. P. (2019). A transshipment model of Seven-Up Bottling Company, Benin Plant, Nigeria. *American Journal of Operations Research*, 9(3), 129–145.
7. Aishwarya, R. (2021). Multi-objective multi item fuzzy trans shipment problem with congestion charge. *Natural Volatiles & Essential Oils*, 8(4), 9935–9956.
8. Bris, M. (2010). Transshipment model in the function Of cost minimization in a logistics system. *Interdisciplinary Management*, 6, 48–59.
9. Baldi, M. M. (2012). The capacitated transshipment location problem under uncertainty: A computational study. *Procedia: Social and Behavioral Sciences*, 39, 424–436.
10. Crainic, G. (2021). The synchronized location-transshipment problem. *Transportation Research Procedia*, 52, 43–50.
11. Ellaimony, E., Ahmed, I., Abdelwahed, K., Ahmed, R., & Khalil, M. (2021). Solving the transshipment problem with fuzzy cost coefficients. *International Journal of Scientific Advances (IJSCIA)*, 2(3), 236–240.
12. Fleischer, L. K. (2001). Faster algorithms for the quickest transshipment problem. *SIAM Journal on Optimization*, 12(1), 18–35.
13. Gani, A. N. (2011). Transshipment problem in fuzzy environment. *International Journal of Mathematical Sciences and Engineering Applications (IJMSEA)*, 5(III), 57–74.
14. Gani, A. N., & Baskaran, R. (2014). Improved Vogel's approximation method to solve fuzzy transshipment problem. *International Journal of Fuzzy Mathematical Archive*, 4(2), 80–87.
15. Gayathri, P., Kannan, K., & Sarala, D. (2015). Max-min method to solve fuzzy transshipment problem. *Applied Mathematical Sciences*, 9(7), 337–343.
16. Gayathri, P., & Subramanian, K. R. (2016). An algorithm to solve fuzzy trapezoidal transshipment problem. *International Journal of Systems Science and Applied Mathematics*, 1(4), 58–62.
17. Giusti, R., Manerba, D., & Tadei, R. (2020). Multiperiod transshipment location–allocation problem with flow synchronization under stochastic handling operations. *Special Issue on Freight Transportation and Logistics*, 78(1), 88–104.
18. Garg, H., Mahmoodirad, A., & Niroomand, S. (2021). Fractional two-stage transshipment problem under uncertainty: Application of the extension principle approach. *Complex and Intelligent Systems*, 7, 807–822.
19. Hurt, V. G., & Tramel, T. E. (1965). Alternative formulations of the transshipment problem. *Journal of Farm Economics*, 47(3), 763–773.
20. Hoadley, B., & Heyman, D. P. (1977). A two-echelon inventory model with purchases, dispositions, shipments, returns and transshipments. *Naval Research Logistics Quarterly*, 24(1), 1–19.
21. Hoppe, B., & Tardos, E. (1995). The quickest transshipment problem. *Mathematics of Operations Research*, 25(1), 512–521.
22. Herer, Y. T., & Rashit, A. (1999). Lateral stock transshipments in a two-location inventory system with fixed and joint replenishment costs. *Naval Research Logistics (NRL)*, 46(5), 525–547.
23. Herer, Y. T., & Tzur, M. (2001). The dynamic transshipment problem. *Naval Research Logistics*, 48(5), 386–408.
24. Herer, Y. T., Tzur, M. & Yücesan, E. (2002). Transshipments: An emerging inventory recourse to achieve supply chain leagility. *International Journal of Production Economics*, 80(3), 201–212.
25. Hu, J., Watson, E., & Schneider, H. (2005). Approximate solutions for multi-location inventory systems with transshipments. *International Journal of Production Economics*, 97(1), 31–43.

26. Hmiden, M., Said, L. B., & Khaled, G. (2009). Transshipment problem with uncertain customer demands and transfer lead time. *International Conference on Computers and Industrial Engineering*, pp. 476–481. France.

27. Hamoud, A. A. (2018). Optimal solution of fuzzy transshipment problem using generalized hexagonal fuzzy numbers. *International Journal of Engineering and Technology*, 7(4), 558–561.

28. Islam, S., & Ray, P. (2017). Multi-objective transhipment problem with an additional entropy objective function using fuzzy programming. *Journal of Fuzzy Set Valued Analysis*, 2017(1), 20–37.

29. Javaid, S., & Gupta, S. N. (2011). A capacitated stochastic linear transshipment problem with prohibited routes. *OPSEARCH*, 48, 30–43.

30. King, G. A., & Logan, S. H. (1964). Optimum location, number, and size of processing plants with raw product and final product shipments. *Journal of Farm Economics*, 46(1), 94–108.

31. Keskin, B. B., & Üste, H. (2007). A scatter search-based heuristic to locate capacitated transshipment points. *Computers and Operations Research*, 34(10), 3112–3125.

32. Khurana, A., & Arora, S. R. (2013). Solving transshipment problems with mixed constraints. *International Journal of Management Science and Engineering Management*, 6(4), 292–297.

33. Kumar, A., Kaur, A., & Gupta, A. (2011). Fuzzy linear programming approach for solving fuzzy transportation problems with transshipment. *Journal of Mathematical Modelling and Algorithms*, 10, 163–180.

34. Khurana, A., & Verma, T. (2015). On a class of capacitated transshipment problems with bounds on rim conditions. *International Journal of Mathematics in Operational Research*, 7(3), 251–280.

35. Kumari, N., & Kumar, R. (2017). Max-min method for solving transshipment problem with mixed constraints. *Global Journal of Pure and Applied Mathematics*, 13(12), 8373–8386.

36. Kumar, A., Chopra, R., & Saxena, R. R. (2020). An efficient algorithm to solve transshipment problem in an uncertain environment. *International Journal of Fuzzy Systems*, 22, 2613–2624.

37. Kumar, A., Chopra, R., & Saxena, R. R. (2022). An enumeration technique for transshipment problem in neutrosophic environment. *Neutrosophic Sets and Systems*, 50, 552–563.

38. Lee, H. L. (1987). A multi-echelon inventory model for repairable items with emergency lateral transshipments. *Management Science*, 33(10), 1302–1316.

39. Lee, Y. H., Jung, J. W., & Jeon, Y. S. (2007). An effective lateral transshipment policy to improve service level in the supply chain. *International Journal of Production Economics*, 106(1), 115–126.

40. Meng, Q., Wang, S., Andersson, H., & Thun, K. (2014). Containership routing and scheduling in liner shipping: Overview and future research directions. *Transportation Science*, 48(2), 265–280.

41. Mohanapriya, S., & Jeyanthi, V. (2017). Modified procedure to solve fuzzy transshipment problem by using trapezoidal fuzzy number. *International Journal of Mathematics and Statistics Invention (IJMSI)*, 4(6), 2321–4759.

42. Mohanaselvi, S., & Prabakaran, K. (2020). Variants of uncertain transshipment problem. *Journal of Mechanical and Production Engineering Research and Development (IJMPERD)*, 10, 6091–6104.

43. Needham, P. M., & Evers, P. T. (1998). The influence of individual cost factors on the use of emergency transshipments. *Transportation Research Part E: Logistics and Transportation Review*, 34(2), 149–160.

44. Notteboom, T., & Rodrigue, J. P. (2008). Containerisation, box logistics and global supply chains: The integration of ports and liner shipping networks. *International Journal of Maritime Economics*, 10, 152–174.

45. Noham, R., & Tzur, M. (2014). The single and multi-item transshipment problem with fixed transshipment costs. *Naval Research Logistics (NRL)*, 61(8), 637–664.

46. Orden, A. (1956). Transshipment problem. *Management Science*, 2(3), 276–285.

47. Perincherry, V., & Kikuchi, S. (1990). A fuzzy approach to the transshipment problem. *Proceedings of the First International Symposium on Uncertainty Modeling and Analysis*, College Park, MD, USA, pp. 330–335.

48. Prabha, S. K., & Vimala, S. (2016). Fuzzy transshipment problem via the method of magnitude. *International Journal of Mathematics and Computer Applications Research*, 6(6), 87–96.

49. Praveena, U. L., & Vinotha, J. M. (2023). A new approach for multi-objective transshipment problem. *European Chemical Bulletin*, 12(S2), 2709–2720.

50. Rajendran, P., & Pandian, P. (2012). Solving fully interval transshipment problems. *International Mathematical Forum*, 7(41), 2027–2035.

51. Reinhardt, L. B., & Pisinger, D. (2012). A branch and cut algorithm for the container shipping network design problem. *Flexible Services and Manufacturing Journal*, 24, 349–374.

52. Sarker, R., & Namatame, A. (2004). Solving transshipment problem using EAs: How good are the solutions. *IEEJ Transactions on Electrical and Electronic Engineering*, 124(10), 1964–1971.
53. Said, L. B., Hmiden, M., & Ghedira, K. (2010). A two-step transshipment model with fuzzy demands and service level constraints. *International Journal of Simulation Modelling*, 9(1), 40–52.
54. Sirajudeen, A., & Gani, N. (2014). A new approach on solving intuitionistic fuzzy transshipment problem. *International Journal of Applied Engineering Research*, 9, 9509–9518.
55. Sujatha, L. (2020). Fuzzy transshipment model using fuzzy one point method. *Advances in Mathematics: Scientific Journal*, 9(8), 6153–6167.
56. Sultan, A. T. A., & Alsaber, A. (2022). Solving vehicle transshipment problem using multi-objective optimization. *Far East Journal of Applied Mathematics*, 114, 65–82.
57. Su, E., Qin, H., Li, J., & Pan, K. (2023). An exact algorithm for the pickup and delivery problem with crowdsourced bids and transshipment. *Transportation Research Part B: Methodological*, 177, 102831.
58. Sarkar, D. D., Kar, S., Basu, K., & Sharma, S. (2023). Solution of a transshipment problem with uncertain parameters under impaired and enhanced flow. *The Journal of Analysis*, 32, 1–27.
59. Skutella, M., (2023). A note on the quickest minimum cost transshipment problem. *Operations Research Letters*, 51(3), 255–258.
60. Thiagarajan, K., & Kandasamy, G. (2018). Fuzzy multi-objective transportation problem: Evolutionary algorithm approach. *Journal of Physics Conference Series*, 1000(1), 012004.
61. Vidhya, V., & Ganesan, K. (2019). Solution of fully fuzzy transshipment problem through a new method. *AIP Conference Proceedings*, 2112(1), 1–5.
62. Vu, H. T. T., & Ko, J. (2022). Integrated inventory transshipment and missing-data treatment using improved imputation-level adjustment for efficient cross-filling. *Sustainability*, 14(19), 12934.
63. Wichapa, N., & Khokhajaikiat, P. (2019). A novel holistic approach for solving the multi-criteria transshipment problem for infectious waste management. *Decision Science Letters*, 8, 441–454.
64. Xu, X., & Wei, Z. (2023). Dynamic pickup and delivery problem with transshipments and LIFO constraints. *Computers and Industrial Engineering*, 175, 108835.
65. Zhou, Z., & Wang, X. (2023). Production, manufacturing, transportation and logistics replenishment and transshipment in periodic-review systems with a fixed order cost. *European Journal of Operational Research*, 307(3), 1240–1247.

7 The Analysis of Sustainability of Women Entrepreneurship by the Use of Complex Fuzzy Matrices

K. T. Divya Mol and Lovelymol Sebastian

7.1 INTRODUCTION

Zadeh [7] was the founder of the concept of fuzzy set theory in 1965. The most important and interesting areas of applications of the theory of fuzzy set are the field of medicine and treatment. Fuzzy matrices are now very rich topic in modeling situations including uncertainty occurred in science, treatments, medical diagnosis, automata, etc.

Fuzzy matrices were introduced by Thomson [5] in 1977, and these concepts were developed by Kim and Roush [3]. K. T. Atanassov [1] was the founder of the notion intuitionistic fuzzy sets, which provides a flexible model to elaborate uncertainty and vagueness in decision-making problems. The concept of fuzzy matrices is used in almost all branches of Science. Fuzzy matrices are a better implement for modeling different problems occurring in uncertain situations in different fields of Science like computer science, robotics, medical science, artificial intelligence, and may others.

In 2002 , Ramot et al. [4] defined the complex fuzzy sets as a generalization of fuzzy sets whose codomain is not restricted to $[0,1]$ but it is expanded on the unit disk in the complex plane (the set of all complex numbers with modulus less than or equal to 1).

Ramot [4] used the belief of complex degree of club in polar coordinates, where the amplitude value is the grade of membership of the object in the complicated fuzzy set and the role of phase is to add additional information related to spatial or temporal periodicity of the object in that precise fuzzy set. The idea of complicated fuzzy matrices was delivered with the aid of Zhi-Quig Zaho and Shong-Quan Ma [8] in 2015. Fuzzy mathematics is a better tool for modeling medical prognosis process and evaluation. By means of using complicated fuzzy matrices, we supplied a new set of rules which will make the analysis more unique.

7.2 PRELIMINARIES

Complex Fuzzy Set: A complex fuzzy set defined on the universe of discourse U is characterized by a membership function $\mu_A(x)$ that assign any element $a \varepsilon A$ a complex valued grade of membership in A.

Using the definition, the values $\mu_A(x)$ lie within the unit disk in the complex plane and are thus of the form $r_A(x)e^{i\omega_A(x)}$, $i = \sqrt{-1}$, $r_A(x)$ and $\omega_A(x)$ are real valued functions, whose values lie in $r_A(x)\varepsilon[0,1]$ and $\omega_A(x)\varepsilon[0,2\pi)$. Thus, the complex fuzzy set can be represented as $A = \{(x,\mu_A(x)) : x\varepsilon U\}$

Remarks: If we do not forget [2], the average sunspot wide variety of the 12 months is 1,800. During the sun activity minimum, handiest few sunspots are there, and for the duration of the solar intensity maximum, there are such a lot of sunspots. By way of the usage of Ramot findings, the complex fuzzy sets are used to provide statistics related to the monthly solar activity and its

DOI: 10.1201/9781003407386-7

role inside the unit circle. The usage of the concept of complex fuzzy sets, the location inside the unit cycle, is represented by means of the segment variable, a real characteristic, and is not a diploma of club. The sun intensity for a selected month is represented through the degree of membership in a fuzzy set. Let us don't forget an example, the complicated fuzzy set prompted through the declaration, "excessive Solar intensity". As we mentioned, if the cycle length is 11 years, it starts from a sun minimal circulate through a sun most and give up at the next solar minimum. Allow zero be the preliminary point, permit π be the solar most, and 2π be the subsequent sun minimal. Suppose the traditional grade of membership of a specific month M_i inside the set with "excessive sun interest" is 0.6 and we assumed that the specific month is at the peak of solar activity for the cycle, then the complicated grade of membership is denoted by means of $0.6e^{i\pi}$. However, Mj is characterized via the complex grade of membership by $0.6e^{i\frac{\pi}{2}}$, then we can say that the month M_j is in the increasing process of solar activity and it is in medium active.

Complex Fuzzy Matrices: A complex fuzzy matrix is defined as a matrix with all its elements from the unit ball in the complex plane.

A complex fuzzy matrix A of order $m \times n$ is defined as $A = [\langle a_{ij}, a_{ij\mu}\rangle]_{m \times n}$, where $a_{ij\mu} = r_A(x)e^{i\omega_A(x)}$, where $r_A(x)\varepsilon[0,1]$ and $\omega_A(x)\varepsilon[0,2\pi)$.

Relative Values: We can say that, the relative values depend on other values. Here we use the following equation to find the relative values from the known values.

$$r_{ij} = (f(\frac{P_i}{d_j}))_{i=1,2,3,4,5,\dots\ j=1,2,3,4,5,\dots}$$

Comparison Matrix: A comparison matrix helps to compare attributes and characteristics of items and helps us to conclude the comparative and relative study. Here we use the following method to create the comparison matrix using the relative values.

$$\text{Comparison matrix, } R = [r_{ij}]; r_{ij} = (f(\frac{P_i}{d_j}))_{i=1,2,3,4,5,\dots\ j=1,2,3,4,5,\dots}$$

7.3 ANALYSIS OF SUSTAINABILITY OF WOMEN ENTREPRENEURSHIP USING COMPLEX FUZZY MATRICES

Women have played an important role in the economic progress all over the world. They are best contributors and supporters of the economy in various ways of working in different sectors. There are many women who run successful businesses. Entrepreneurship skills in various women are different, and some of them may be more successful than the others. The known fact is that many of the world's most successful and large-scale enterprises are owned and run by women.

In today's world, women are recognized as most successful entrepreneurs because of their strong desire, qualities, and skills for strong economic development. Due to such an important contribution of women in economic development, it is more desirable to study the factors affecting the sustainability of women entrepreneurs.

The analysis has a lot of confounding alternatives. Therefore, the analysis procedure is based on the study of combinations of variables with complex outcomes. At first, we consider the input variable with their membership functions. In the second step, we consider the use of the notions complex fuzzy sets and complex fuzzy matrices to identify the major factor affecting entrepreneurship.

First we will consider some main factors that affect women entrepreneurship as the input variables.

Table 7.1

Age Group: Input Variable Data Table

.	d_1	d_2	d_3	d_4	d_5	d_6	d_7	d_8
A_1	26	14	20	3	22	14	12	10
A_2	10	12	9	23	29	11	13	12
A_3	16	22	24	8	23	29	17	15
A_4	22	8	13	17	25	24	13	9
A_5	7	19	21	6	17	12	29	22
A_6	17	12	16	10	10	9	25	11
A_7	24	4	18	18	10	7	27	24
A_8	15	12	21	28	18	6	5	14

So, the input variables are [6]:-

1. Efficiency in mobilizing funds - d_1
2. Optimistic approach toward criticism - d_2
3. Self-efficiency and self-confidence - d_3
4. Attraction toward idea of networking - d_4
5. Need for achievements - d_5
6. Family support - d_6
7. Adequate knowledge and skills - d_7
8. Risk-taking attitude - d_8

While considering these input variables, we assign some range for defining membership values for these input variables (Table 7.1).

For that, we have taken a survey of 240 women of different age groups (taking 30 from each age groups) to get the data for our analysis.

The eight different age groups are,

A_1 : $18 - 21$
A_2 : $22 - 25$
A_3 : $26 - 29$
A_4 : $30 - 33$
A_5 : $34 - 37$
A_6 : $38 - 41$
A_7 : $42 - 45$
A_8 : $46 - 49$

The membership function for each of the input variables defined as follows:

7.3.1 EFFICIENCY IN MOBILIZING FUNDS - d_1

The d_1 variable stands for the efficiency in mobilizing funds. From the survey, we can say that the range for input variables is between 0 and 30.

When d_1 ranges $0 \leq x < 10$, we use the phase value 0 to represent the range of d_1 in the fuzzy set.

When d_1 range is $10 \leq x \leq 20$, we use the phase value $\frac{2\pi}{3}$ to represent the particular range of d_1 in the fuzzy set.

When d_1 range $20 < x \le 30$, we use the phase value $\frac{4\pi}{3}$ to represent the particular range of d_1 in the fuzzy set.

The membership functions is

$$\mu_{d_1}(x) = \begin{cases} 0 & 0 \le x \le 10 \\ \frac{20-x}{10} & 10 < x < 20 \\ 1 & 20 \le x \le 30 \end{cases}$$

$$= \begin{cases} 0e^{i0} \\ \frac{20-x}{10} e^{i\frac{2\pi}{3}} \\ 1e^{i\frac{4\pi}{3}} \end{cases}$$

Similarly, all other input variable range is 0–30, so we can use the same membership function.

7.3.1.1 Remark

In all the membership functions of each fuzzy sets mentioned above, we introduced the phase value to represent the position of that particular input variable in that fuzzy set along with its degree of membership. So that all the above defined fuzzy sets become complex fuzzy sets. This aspect helps us to use the concept of complex fuzzy matrices, which is very helpful in the further developments of decision support system.

Here for the age group A_1, the value of d_1 is 26, then the membership value becomes $\mu_{d_1} = 1e^{i\frac{4\pi}{3}}$, obtained by substituting the value 26 in the corresponding membership function.

Similarly,

$$\begin{aligned}
d_2 &= 14 \longrightarrow \mu_{d_2} = 0.6e^{i\frac{2\pi}{3}} \\
d_3 &= 20 \implies \mu_{d_3} = 1e^{i\frac{4\pi}{3}} \\
d_4 &= 3 \implies \mu_{d_4} = 0e^{i0} \\
d_5 &= 22 \implies \mu_{d_5} = 1e^{i\frac{4\pi}{3}} \\
d_6 &= 14 \implies \mu_{d_6} = 0.6e^{i\frac{2\pi}{3}} \\
d_7 &= 12 \implies \mu_{d_7} = 0.8e^{i\frac{2\pi}{3}} \\
d_8 &= 10 \implies d_8 = 0e^{i0}
\end{aligned}$$

Then, the first row matrix for the age group A_1 is

$$A_1 \longrightarrow \begin{bmatrix} 1e^{i\frac{4\pi}{3}} \\ 0.6e^{i\frac{2\pi}{3}} \\ 1e^{i\frac{4\pi}{3}} \\ 0e^{i0} \\ 1e^{i\frac{4\pi}{3}} \\ 0.6e^{i\frac{2\pi}{3}} \\ 0.8e^{i\frac{2\pi}{3}} \\ 0e^{i0} \end{bmatrix}$$

Suppose that for the age group A_2, the value of d_1 is 10, then the membership value becomes $\mu_{d_1} = 0e^{i0}$, obtained by substituting the value 10 in the corresponding membership function.

Similarly,

$$
\begin{aligned}
d_2 = 12 &\longrightarrow \mu_{d_2} = 0.8e^{i\frac{2\pi}{3}} \\
d_3 = 9 &\implies \mu_{d_3} = 0e^{i0} \\
d_4 = 23 &\implies \mu_{d_4} = 1e^{i\frac{4\pi}{3}} \\
d_5 = 29 &\implies \mu_{d_5} = 1e^{i\frac{4\pi}{3}} \\
d_6 = 11 &\implies \mu_{d_6} = 0.9e^{i\frac{2\pi}{3}} \\
d_7 = 13 &\implies \mu_{d_7} = 0.7e^{i\frac{2\pi}{3}} \\
d_8 = 12 &\implies d_8 = 0.8e^{i\frac{2\pi}{3}}
\end{aligned}
$$

Then, the first row matrix for the age group A_2 is

$$
A_2 \longrightarrow
\begin{bmatrix}
0e^{i0} \\
0.8e^{i\frac{2\pi}{3}} \\
0e^{i0} \\
1e^{i\frac{4\pi}{3}} \\
1e^{i\frac{4\pi}{3}} \\
0.9e^{i\frac{2\pi}{3}} \\
0.7e^{i\frac{2\pi}{3}} \\
0.8e^{i\frac{2\pi}{3}}
\end{bmatrix}
$$

Here for the age group A_3, the value of d_1 is 16, then the membership value becomes $\mu_{d_1} = 0.4e^{i\frac{2\pi}{3}}$, obtained by substituting the value 16 in the corresponding membership function.

Similarly,

$$
\begin{aligned}
d_2 = 22 &\longrightarrow \mu_{d_2} = 1e^{i\frac{4\pi}{3}} \\
d_3 = 24 &\implies \mu_{d_3} = 1e^{i\frac{4\pi}{3}} \\
d_4 = 8 &\implies \mu_{d_4} = 0e^{i0} \\
d_5 = 23 &\implies \mu_{d_5} = 1e^{i\frac{4\pi}{3}} \\
d_6 = 25 &\implies \mu_{d_6} = 1e^{i\frac{4\pi}{3}} \\
d_7 = 17 &\implies \mu_{d_7} = 0.3e^{i\frac{2\pi}{3}} \\
d_8 = 15 &\implies d_8 = 0.5e^{i\frac{2\pi}{3}}
\end{aligned}
$$

Then, the first row matrix for the age group A_3 is

$$
A_3 \longrightarrow
\begin{bmatrix}
0.4e^{i\frac{2\pi}{3}} \\
1e^{i\frac{4\pi}{3}} \\
1e^{i\frac{4\pi}{3}} \\
0e^{i0} \\
1e^{i\frac{4\pi}{3}} \\
1e^{i\frac{4\pi}{3}} \\
0.3e^{i\frac{2\pi}{3}} \\
0.5e^{i\frac{4\pi}{3}}
\end{bmatrix}
$$

Here for the age group A_4, the value of d_1 is 22, then the membership value becomes $\mu_{d_1} = 1e^{i\frac{4\pi}{3}}$, obtained by substituting the value 22 in the corresponding membership function.

Similarly,

$$d_2 = 8 \longrightarrow \mu_{d_2} = 0e^{i0}$$
$$d_3 = 13 \implies \mu_{d_3} = 0.7e^{i\frac{2\pi}{3}}$$
$$d_4 = 17 \implies \mu_{d_4} = 0.3e^{i\frac{2\pi}{3}}$$
$$d_5 = 25 \implies \mu_{d_5} = 1e^{i\frac{4\pi}{3}}$$
$$d_6 = 24 \implies \mu_{d_6} = 1e^{i\frac{4\pi}{3}}$$
$$d_7 = 13 \implies \mu_{d_7} = 0.7e^{i\frac{2\pi}{3}}$$
$$d_8 = 9 \implies d_8 = 0e^{i0}$$

Then, the first row matrix for the age group A_4 is

$$A_4 \longrightarrow \begin{bmatrix} 1e^{i\frac{4\pi}{3}} \\ 0e^{i0} \\ 0.7e^{i\frac{2\pi}{3}} \\ 0.3e^{i\frac{2\pi}{3}} \\ 1e^{i\frac{4\pi}{3}} \\ 1e^{i\frac{4\pi}{3}} \\ 0.7e^{i\frac{2\pi}{3}} \\ 0e^{i0} \end{bmatrix}$$

Here for the age group A_5, the value of d_1 is 7, then the membership value becomes $\mu_{d_1} = 0e^{i0}$, obtained by substituting the value 7 in the corresponding membership function.

Similarly,

$$d_2 = 19 \longrightarrow \mu_{d_2} = 0.1e^{i\frac{2\pi}{3}}$$
$$d_3 = 21 \implies \mu_{d_3} = 1e^{i\frac{4\pi}{3}}$$
$$d_4 = 6 \implies \mu_{d_4} = 0e^{i0}$$
$$d_5 = 17 \implies \mu_{d_5} = 0.3e^{i\frac{2\pi}{3}}$$
$$d_6 = 12 \implies \mu_{d_6} = 0.8e^{i\frac{2\pi}{3}}$$
$$d_7 = 29 \implies \mu_{d_7} = 1e^{i\frac{4\pi}{3}}$$
$$d_8 = 22 \implies d_8 = 1e^{i\frac{4\pi}{3}}$$

Then, the first row matrix for the age group A_5 is

$$A_5 \longrightarrow \begin{bmatrix} 0e^{i0} \\ 0.1e^{i\frac{2\pi}{3}} \\ 1e^{i\frac{4\pi}{3}} \\ 0e^{i0} \\ 0.3e^{i\frac{2\pi}{3}} \\ 0.8e^{i\frac{2\pi}{3}} \\ 1e^{i\frac{4\pi}{3}} \\ 1e^{i\frac{4\pi}{3}} \end{bmatrix}$$

Here for the age group A_6, the value of d_1 is 17, then the membership value becomes $\mu_{d_1} = 0.3e^{i\frac{2\pi}{3}}$, obtained by substituting the value 17 in the corresponding membership function.

Similarly,

$$
\begin{aligned}
d_2 &= 12 \longrightarrow \mu_{d_2} = 0.9e^{i\frac{2\pi}{3}} \\
d_3 &= 16 \implies \mu_{d_3} = 0.4e^{i\frac{2\pi}{3}} \\
d_4 &= 10 \implies \mu_{d_4} = 0e^{i0} \\
d_5 &= 10 \implies \mu_{d_5} = 0e^{i0} \\
d_6 &= 9 \implies \mu_{d_6} = 0e^{i0} \\
d_7 &= 25 \implies \mu_{d_7} = 1e^{i\frac{4\pi}{3}} \\
d_8 &= 11 \implies d_8 = 0.9e^{i\frac{2\pi}{3}}
\end{aligned}
$$

Then, the first row matrix for the age group A_6 is

$$
A_6 \longrightarrow
\begin{bmatrix}
0.3e^{i\frac{2\pi}{3}} \\
0.8e^{i\frac{2\pi}{3}} \\
0.4e^{i\frac{2\pi}{3}} \\
0e^{i0} \\
0e^{i0} \\
0e^{i0} \\
1e^{i\frac{4\pi}{3}} \\
0.9e^{i\frac{2\pi}{3}}
\end{bmatrix}
$$

Here for the age group A_7, the value of d_1 is 24, then the membership value becomes $\mu_{d_1} = 1e^{i\frac{4\pi}{3}}$, obtained by substituting the value 24 in the corresponding membership function.

Similarly,

$$
\begin{aligned}
d_2 &= 4 \longrightarrow \mu_{d_2} = 0e^{i0} \\
d_3 &= 18 \implies \mu_{d_3} = 0.2e^{i\frac{2\pi}{3}} \\
d_4 &= 18 \implies \mu_{d_4} = 0.2e^{i\frac{2\pi}{3}} \\
d_5 &= 10 \implies \mu_{d_5} = 0e^{i0} \\
d_6 &= 7 \implies \mu_{d_6} = 0e^{i0} \\
d_7 &= 27 \implies \mu_{d_7} = 1e^{i\frac{4\pi}{3}} \\
d_8 &= 24 \implies d_8 = 1e^{i\frac{4\pi}{3}}
\end{aligned}
$$

Then, the first row matrix for the age group A_7 is

$$
A_7 \longrightarrow
\begin{bmatrix}
1e^{i\frac{4\pi}{3}} \\
0e^{i0} \\
0.2e^{i\frac{2\pi}{3}} \\
0.2e^{i\frac{2\pi}{3}} \\
0e^{i0} \\
0e^{i0} \\
1e^{i\frac{4\pi}{3}} \\
1e^{i\frac{4\pi}{3}}
\end{bmatrix}
$$

Here for the age group A_8, the value of d_1 is 15, then the membership value becomes $\mu_{d_1} = 0.5e^{i\frac{2\pi}{3}}$, obtained by substituting the value 10 in the corresponding membership function.

Similarly,

$$d_2 = 12 \longrightarrow \mu_{d_2} = 0.8e^{i\frac{2\pi}{3}}$$
$$d_3 = 21 \implies \mu_{d_3} = 1e^{i\frac{4\pi}{3}}$$
$$d_4 = 28 \implies \mu_{d_4} = 1e^{i\frac{4\pi}{3}}$$
$$d_5 = 18 \implies \mu_{d_5} = 0.2e^{i\frac{2\pi}{3}}$$
$$d_6 = 6 \implies \mu_{d_6} = 0e^{i0}$$
$$d_7 = 5 \implies \mu_{d_7} = 0e^{i0}$$
$$d_8 = 14 \implies d_8 = 0.6e^{i\frac{2\pi}{3}}$$

Then, the first row matrix for the age group A_8 is

$$A_8 \longrightarrow \begin{bmatrix} 0.5e^{i\frac{2\pi}{3}} \\ 0.8e^{i\frac{2\pi}{3}} \\ 1e^{i\frac{4\pi}{3}} \\ 1e^{i\frac{4\pi}{3}} \\ 0.2e^{i\frac{2\pi}{3}} \\ 0e^{i0} \\ 0e^{i0} \\ 0.6e^{i\frac{2\pi}{3}} \end{bmatrix}$$

So, the age group-causes of input variable complex fuzzy matrix is given by

$$\begin{bmatrix} 1e^{i\frac{4\pi}{3}} & 0.6e^{i\frac{2\pi}{3}} & 1e^{i\frac{4\pi}{3}} & 0e^{i0} & 1e^{i\frac{4\pi}{3}} & 0.6e^{i\frac{2\pi}{3}} & 0.8e^{i\frac{2\pi}{3}} & 0e^{i0} \\ 0e^{i0} & 0.8e^{i\frac{2\pi}{3}} & 0e^{i0} & 1e^{i\frac{4\pi}{3}} & 1e^{i\frac{4\pi}{3}} & 0.9e^{i\frac{2\pi}{3}} & 0.7e^{i\frac{2\pi}{3}} & 0.8e^{i\frac{2\pi}{3}} \\ 0.4e^{i\frac{2\pi}{3}} & 1e^{i\frac{4\pi}{3}} & 1e^{i\frac{4\pi}{3}} & 0e^{i0} & 1e^{i\frac{4\pi}{3}} & 1e^{i\frac{4\pi}{3}} & 0.3e^{i\frac{2\pi}{3}} & 0.5e^{i\frac{4\pi}{3}} \\ 1e^{i\frac{4\pi}{3}} & 0e^{i0} & 0.7e^{i\frac{2\pi}{3}} & 0.3e^{i\frac{2\pi}{3}} & 1e^{i\frac{4\pi}{3}} & 1e^{i\frac{4\pi}{3}} & 0.7e^{i\frac{2\pi}{3}} & 0e^{i0} \\ 0e^{i0} & 0.1e^{i\frac{2\pi}{3}} & 1e^{i\frac{4\pi}{3}} & 0e^{i0} & 0.3e^{i\frac{2\pi}{3}} & 0.8e^{i\frac{2\pi}{3}} & 1e^{i\frac{4\pi}{3}} & 1e^{i\frac{4\pi}{3}} \\ 0.3e^{i\frac{2\pi}{3}} & 0.8e^{i\frac{2\pi}{3}} & 0.4e^{i\frac{2\pi}{3}} & 0e^{i0} & 0e^{i0} & 0e^{i0} & 1e^{i\frac{4\pi}{3}} & 0.9e^{i\frac{2\pi}{3}} \\ 1e^{i\frac{4\pi}{3}} & 0e^{i0} & 0.2e^{i\frac{2\pi}{3}} & 0.2e^{i\frac{2\pi}{3}} & 0e^{i0} & 0e^{i0} & 1e^{i\frac{4\pi}{3}} & 1e^{i\frac{4\pi}{3}} \\ 0.5e^{i\frac{2\pi}{3}} & 0.8e^{i\frac{2\pi}{3}} & 1e^{i\frac{4\pi}{3}} & 1e^{i\frac{4\pi}{3}} & 0.2e^{i\frac{2\pi}{3}} & 0e^{i0} & 0e^{i0} & 0.6e^{i\frac{2\pi}{3}} \end{bmatrix}_{8\times 8}$$

Calculate the relative values of $f(\frac{A_i}{d_i})$ and form the comparison matrix $[r_{ij}]$

$$R = (r_{ij}) = \left(f\left(\frac{A_i}{d_j}\right)\right)_{i=1,2,3,4,5,6,7,8.\,,j=1,2,3,4,5,6,7,8}$$

$$r_{11} = f\left(\frac{A_1}{d_1}\right) = \frac{|\mu_{d_1}(A_1)| - |\mu_{A_1}(d_1)|}{\max\{|\mu_{d_1}(A_1)|, |\mu_{A_1}(d_1)|\}}$$
$$= \frac{1-1}{\max\{1,1\}} = \frac{0}{1} = 0$$

$$r_{12} = f\left(\frac{A_1}{d_2}\right) = \frac{|\mu_{d_2}(A_1)| - |\mu_{A1}(d_2)|}{\max\{|\mu_{d_2}(A_1)|, |\mu_{A_1}(d_2)|\}}$$
$$= \frac{0.6-0}{\max\{0.6,0\}} = \frac{0.6}{0.6} = 1$$

$$r_{13} = f\left(\frac{A_1}{d_3}\right) = \frac{|\mu_{d_3}(A_1)| - |\mu_{A_1}(d_3)|}{\max\{|\mu_{d_3}(A_1)|, |\mu_{A_1}(d_3)|\}}$$

$$= \frac{1 - 0.4}{\max\{1, 0.4\}} = \frac{0.6}{1} = 0.6$$

$$r_{14} = f\left(\frac{A_1}{d_4}\right) = \frac{|\mu_{d_4}(A_1)| - |\mu_{A_1}(d_4)|}{\max\{|\mu_{d_4}(A_1)|, |\mu_{A_1}(d_4)|\}}$$

$$= \frac{0 - 1}{\max\{0, 1\}} = \frac{-1}{1} = -1$$

$$r_{15} = f\left(\frac{A_1}{d_5}\right) = \frac{|\mu_{d_5}(A_1)| - |\mu_{A_1}(d_5)|}{\max\{|\mu_{d_5}(A_1)|, |\mu_{A_1}(d_5)|\}}$$

$$= \frac{1 - 0}{\max\{1, 0\}} = \frac{1}{1} = 1$$

$$r_{16} = f\left(\frac{A_1}{d_6}\right) = \frac{|\mu_{d_6}(A_1)| - |\mu_{A_1}(d_6)|}{\max\{|\mu_{d_6}(A_1)|, |\mu_{A_1}(d_6)|\}}$$

$$= \frac{0.6 - 0.3}{\max\{0.6, 0.3\}} = \frac{0.3}{0.6} = 0.5$$

$$r_{17} = f\left(\frac{A_1}{d_7}\right) = \frac{|\mu_{d_7}(A_1)| - |\mu_{A_1}(d_7)|}{\max\{|\mu_{d_7}(A_1)|, |\mu_{A_1}(d_7)|\}}$$

$$= \frac{0.8 - 1}{\max\{0.8, 1\}} = \frac{-0.2}{1} = -0.2$$

$$r_{18} = f\left(\frac{A_1}{d_8}\right) = \frac{|\mu_{d_8}(A_1)| - |\mu_{A_1}(d_8)|}{\max\{|\mu_{d_8}(A_1)|, |\mu_{A_1}(d_8)|\}}$$

$$= \frac{0 - 0.5}{\max\{0, 0.5\}} = \frac{-0.5}{0.5} = -1$$

Similarly,

$r_{21} = -1, r_{22} = 0, r_{23} = -1, r_{24} = 1, r_{25} = 0.9, r_{26} = 0.11, r_{27} = 1, r_{28} = 0$

$r_{31} = -0.6, r_{32} = 1, r_{33} = 0, r_{34} = -1, r_{35} = 0, r_{36} = 0.6, r_{37} = 0.333, r_{38} = -0.5$

$r_{41} = 1; r_{42} = -1; r_{43} = 1; r_{44} = 0; r_{45} = 1; r_{46} = 1; r_{47} = -0.5; r_{48} = -1$

$r_{51} = -1; r_{52} = -0.9; r_{53} = 0; r_{54} = -1; r_{55} = 0; r_{56} = 1; r_{57} = 1; r_{58} = 0.8$

$r_{61} = 0.5; r_{62} = -0.111; r_{63} = -0.6; r_{64} = -1; r_{65} = -1; r_{66} = 0; r_{67} = 1; r_{68} = 0.778$

$r_{71} = 0.2; r_{72} = -1; r_{73} = -0.333; r_{74} = 0.714; r_{75} = -1; r_{76} = 1; r_{77} = 0; r_{78} = 1$

$$r_{81} = 1; \ r_{82} = 0; \ r_{83} = 0.5; \ r_{84} = 1; \ r_{85} = -0.8; \ r_{86} = -1; \ r_{87} = -1; \ r_{88} = 0$$

So, the comparison matrix is,

$$\begin{bmatrix} 0.0000 & 1.0000 & 0.6000 & 0.000 & -1.00 & 1.0000 & 0.5000 & -1.000 \\ -1.000 & 0.0000 & -1.000 & 1.0000 & 0.9000 & 0.1111 & 1.0000 & 0.0000 \\ -0.6000 & 1.0000 & 0.0000 & -1.000 & 0.0000 & 0.6000 & 0.3333 & -0.500 \\ 1.0000 & -1.000 & 1.0000 & 0.000 & 1.0000 & 1.0000 & -0.500 & -1.000 \\ -1.000 & -0.900 & 0.0000 & -1.000 & 0.0000 & 1.0000 & 1.0000 & 0.8000 \\ 0.5000 & -0.111 & -0.600 & -1.000 & -1.000 & 0.0000 & 1.0000 & 0.7778 \\ 0.2000 & -1.0000 & -0.333 & 0.7140 & -1.000 & 1.0000 & 0.000 & 1.0000 \\ 1.0000 & 0.0000 & 0.5000 & 1.0000 & -0.800 & -1.000 & -1.000 & 0.0000 \end{bmatrix}_{8 \times 8}$$

From the ranking of the problem, conclude that the main difficulties faced by each age group can be found by taking maximum in each row, so for the age group A_1 is d_2, d_6, i.e., the optimistic approach toward criticism and family support; for the age group A_2 is d_4, d_7, i.e., the attraction toward the idea of networking and adequate knowledge and skills; for the age group A_3 is d_2, i.e., the optimistic approach toward criticism; for the age group A_4 is d_1, d_3, d_5, d_6, i.e., efficiency in mobilizing funds, self-efficiency and self-confidence, need for achievements and family support; for the age group A_5 is d_6, d_7, i.e., family support and adequate knowledge and skills; for the age group A_7 is d_6, d_8, i.e., family support and risk-taking ability; and for the age group A_8 is d_1, d_4, i.e., the efficiency in mobilizing funds and attraction toward the idea of networking.

7.4 CONCLUSION

Through this chapter, using the concept of complex fuzzy matrices we developed a mathematical model. The notion of complex fuzzy sets has gone through an evolutionary process since they first introduced. Medical field and modeling are the best fields in which the term complex fuzzy sets is applicable. The theory of fuzzy matrices in the field of human diseases diagnosis and decision-making problems was recognized quite early. The doctor generally gets knowledge about the patient from the past history and laboratory test results, and the knowledge provided by each of these factors carries with it varying degrees of uncertainty. This problem can be overcome by using the concept of fuzzy matrices.

Therefore , in this chapter, through the use of the newly introduced concept of complex fuzzy matrices, we identify the key problem factors faced by women entrepreneurs of different ages. The method assumes knowledge of risk factors affecting the sustainability of women entrepreneurs of different ages. This method will help the analyst and the country to identify the main problems faced by women entrepreneurs of a particular age group. That is, if the risk factor of a particular age group is self-efficacy and self-confidence, this method will assist the country to make entrepreneurs of that particular age group more self-efficacy and self-confidence by launching some innovative methods and strategies. With such an approach, we can make women entrepreneurs more sustainable and successful. In the future, this method can be applied to other types of modeling problems and decision-making in the treatment of various diseases.

REFERENCES

1. Atanassov, K. T.: Intuitionistic Fuzzy Sets, *Fuzzy Sets and System*, **20**(1986), 87–96.
2. Geetha, T. and Usha, A.: Circulant Fuzzy Matrix in Human Body Diseases, *International Journal of Pure and Applied Mathematics*, **117**(2017), 99–106.
3. Kim, K. H. and Roush, F. W.: Generalised Fuzzy Matrices, *Fuzzy Sets and Systems*, **4**(1980), 293–315.
4. Ramot, D., Milo, R., Friedman, M., and Kandel, A.: Complex Fuzzy Sets, *IEEE Transactions on Fuzzy Systems*, **10**(2002), 171–186.

5. Thomson, M. G.: Convergence of Powers of Fuzzy Matrix, *Journal of Mathematical Analysis and Applications*, **57**(1997), 476–480.

6. Visalakshi, V., Suvitha, V., and Sruthi, S.: An Analysis on Sustainability of Women Entrepreneurship Using Fuzzy Matrix, *Annals of the Romanian Society for Cell Biology*, **25**(2021), 15320–15326.

7. Zadeh, L. A.: Fuzzy Sets, *Information and Control*, **8**(1965), 338–353.

8. Zaho, Z.-Q. and Ma, S.-Q.: Complex Fuzzy Matrices and Its Convergence Problem Research, *Fuzzy System and Operation Research*, **367**(2015), 157–162.

8 Operational Controllability of Neutral Higher-Order Integrodifferential Systems

B. Radhakrishnan and P. Anukokila

8.1 INTRODUCTION

Diverse branches of research and technology employ differential equations, notably if there is a known or proposed deterministic relationship involving certain continuously fluctuating values and their rates of change in space and/or time. This has been demonstrated by traditional mechanics, which defines the movement of a body as a function of its location and velocity over time. With the aid of Newton's principles, it is possible to dynamically characterize these variables as a differential equation for the structure's uncertain position as an indicator of period. The most common method to represent an issue with an evolution partial differential equation's beginning value or original-boundary value is by a theoretical differential formulation

$$z'(\tau) = g(\tau, z(\tau)) \tag{8.1}$$

under an appropriate function region. The function g defines when the equation acts on the variables in the space while taking boundary circumstances into account when describing the space or domain of g. Although the resemblances among Equation (8.1) and legitimate ordinary differential equation are merely formal, they offer heuristic insight into the issue, pointers for extending inferences from ordinary to partial differential equations, and emphasize unification, which leads to the discovery of common threads and thought economy. The "abstract" approach isn't ideal in every circumstance (many controllability conclusions, for example, are dependent on features of partial differential equations or are difficult to reformulate in the translation to Equation (8.1)), but it's ideal for optimization issues. Many of the strategies are oblivious to the kind of equation (with some fine tuning) and are formally similar to classical algorithms for systems of ordinary differential equations.

A fascinating subject with numerous implications in analysis and other branches of mathematics is the mathematical framework of differential equations in abstract domains. Depending on the nature of the issues, ODEs, functional differential equations, PDEs, and occasionally a mix of interacting systems of ODEs and PDEs can potentially be employed. Many issues in the physical sciences, engineering, biology, and applied mathematics are solved using nonlinear differential and integral equations in abstract spaces. With significant applications in many areas of the research and other disciplines, the mathematical framework of nonlinear differential and integral equations within abstract domains is quickly developing.

Neutral differential equations have been widely used in practical mathematics over the past few decades (see [1, 2–9]). Around certain circumstances, such as the so-called neutral differential equations for difference, the delayed input exists in both the gradient of the state factor and the independent variable. A system with a neutral functional differential equation comprises a single that includes both the system's current state and the derivatives of the system's historical or functional past. The book by Hale and Verduyn Lunel [10] and the sources contained within serve as a valuable companion to the available research for neutral functional differential equations.

DOI: 10.1201/9781003407386-8

Control theory is dependent on the qualitative characteristic of dynamical systems known as controllability. In deterministic system theory, the concept of controllability has played an essential role. The controllability of deterministic equations is well recognized and commonly used in the analysis and design of control systems. Early in the 1960s, a framework of controllability centered around state space descriptions of simultaneous time invariant and time-varying linear systems of control was put forward, kicking off a systematic study of controllability. Any control system is considered to be controllable if some control signals can affect or control every state relating to a process in real time. Controllability, in general, means that a set of admissible controls can transition a dynamical system from any viable past trajectory in system behavior to any conceivable future trajectory after a finite time; that is, there exist systems that are entirely controllable. Other sorts of controllability, including approximate, null, local null, local approximate null controllability, and so on, can be shown if the system is not entirely controllable.

This theory has been extensively investigated over the past several decades with regard to the controllability problems related to finite-dimensional systems that feature ODEs. Only when the algebraic Kalman rank criterion is met, a system is considered controllable around finite dimensions. This characteristic states that once a system has been managed for a while, it is always controllable. This reflects no longer true for systems with infinite dimensions that are described by PDEs. In particular, in the frame work of the wave equation, a model in which propagation occurs with finite velocity, the control time must be large enough for the influence of the control to reach everywhere in order for controllability qualities to be true. When physical issues are simulated, semi-linear evolution equations are frequently used as the model. A semi-linear evolution system in a Banach space can be used to simulate such challenges in control fluid flow. In Ref. [11], the control challenges that rise to this type of model, as well as the model equation that results, are discussed.

In order to answer the question of whether linear systems characterized by differential equations throughout Banach spaces are controllable, the techniques used by the authors in Refs. [12, 13] can be applied. Nonlinear systems' controllability has recently gained a lot of attention, and a lot of articles have begun to appear. In finite-dimensional domains, fixed point principles [3, 14–16] ensure the controllability of nonlinear systems described by ODEs. Semi-group theory [17] is being used to extend this strategy to infinite-dimensional spaces. Many authors have used fixed point notions to examine the controllability of nonlinear systems with various types of non-linearity [18, 19].

Because of its various applications, the study of second-order linear and nonlinear differential equations in abstract spaces has arisen as a separate topic of modern research in recent years. A lot of research has been done on the existence problems associated with second-order differential equations utilizing the idea of highly continuously cosine families of bounded linear operators [20–22]. Fattorini's work on cosine families is the most fundamental and comprehensive [23, 24]. In many circumstances, treating second-order abstract differential equations directly rather than converting them to first-order systems is preferable. The theory of highly continuous cosine family theory [27–29, 45] is a handy tool for studying an abstract second-order equation. Numerous authors have studied the controllability issue for the second-order linear and nonlinear differential and integrodifferential structures within Banach spaces using the fixed point theorems [25, 26].

Therefore, it is vital to understand the controllability issue associated with such structures in Banach spaces. Because of its various applications, the study of second-order integrodifferential equations in abstract spaces has arisen as a separate subject of current research in recent years. Second-order differential equations, also known as integrodifferential equations, are used to simulate a number of physical processes, including the vibration of pivoted bars, the axial motion of an extensible beam, and more. Consequently, it is essential to research the ability to control issue for such systems in Banach spaces.

8.2　MOTIVATION

Since this chapter's purpose is to examine the controllability with nonlinear differential and integrodifferential systems, we begin by describing how these systems appear in many fields of study.

8.2.1　BALL JUMPING ON A HORIZONTAL SURFACE [30]

We consider a ball that is jumping on a flat horizontal surface. Loss of energy, caused by the friction of the surface, is characterized by a constant μ. This process is simulated by a second-order differential equation of the form

$$m\frac{d^2x}{dt^2} = F$$

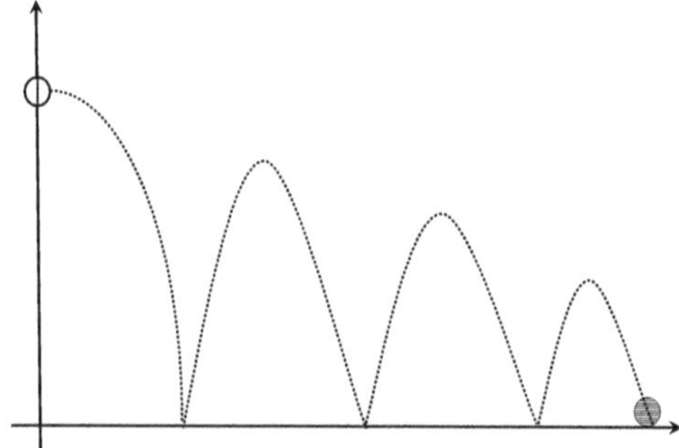

where m is the mass of the ball, $F = -mg$ is the force, and g is the acceleration of Earth's gravitation. At any time when the ball touches the surface, the vertical component of the vector velocity changes sign. The state of this process is described by the vertical position, velocity, and horizontal position of the ball. This process can be described with the impulsive differential equation

$$\frac{dx_1}{dt} = x, \quad \frac{dx_2}{dt} = -g, \quad \frac{dx_3}{dt} = v_0,$$

with initial condition $x = (x_1, x_2, x_3) = (h_0, 0, 0)$ and with condition of jump

$$I_k(x_1, x_2, x_3) = (x_1, -\mu \cdot x_2, x_3) \text{ for } u(x) = x_1 = 0.$$

In this model, x_1 is the vertical position, x_2 is the velocity, and x_3 is the horizontal position of the ball.

8.2.2　THEORY OF ELASTIC STRING

Consider the following integrodifferential equation

$$\frac{\partial^2 z(y,\tau)}{\partial t^2} + c(\tau)\frac{\partial z(y,\tau)}{\partial t} \quad - \quad M\left(\int_{-\infty}^{+\infty} |\frac{\partial z(y,s)}{\partial s}|^2 ds\right)\frac{\partial^2 z(y,\tau)}{\partial y^2} + z(y,\tau)$$

$$= l(\tau, y, z(\tau, y)), \ 0 \le \tau < \infty, \ y \in \mathbb{R},$$

$$z(0, y) = z_0(y),$$

$$\frac{\partial z(y,0)}{\partial \tau} = z_1(y), \ y \in \mathbb{R}.$$

This sort of equation may be found in the research on the nonlinear behavior of a rigid flexible string, according to Ref. [31]. The fundamental physical presumptions included based on the string's longitudinal strain are relatively modest while the tension $\mathscr{F}$ is constant across the string, though it may fluctuate over time to account for variations in the string's arc length. The presumption that F relies on the string's arc length $\mathscr{S}$ at time $\tau \leq 0$ by the relation $\mathscr{F} = \mathscr{F}_0 + D[(\mathscr{S} - \mathscr{L})/\mathscr{L}]$ leads to non-linearity where $\mathscr{F}_0$ is the minimum tension, $\mathscr{L}$ is the minimum length, and D is a physical constant.

The investigation associated with dynamical buckling that occurs in a hinged stretchable beam that is either stretched or compressed by an axial force involves various sorts of integrodifferential equations that are comparable to the ones mentioned previously. An equation of the following form

$$
\begin{aligned}
w_\tau(\tau,y) + \Psi(w(\tau,y))_y &= \int_0^t b(t-s)\Psi(w(s,y))_x ds + g(\tau,y), \ \tau \in [0,b], y \in \mathbb{R}, \\
w(0,y) &= \phi(y), \ y \in \mathbb{R},
\end{aligned}
$$

takes place with a memory-based nonlinear conversion rule.

The above type of equation can be formulated abstractly as

$$
\begin{aligned}
\dot{z}(\tau) + \mathscr{A}(\tau,z(\tau))z(\tau) &= \mathscr{B}v(\tau) + g\left(\tau,z,\int_0^\tau h(\tau,s,z(s))ds\right), \ \tau \in [0,b], \\
z(0) &= z_0,
\end{aligned}
$$

where g,h are provided nonlinear functions and $-\mathscr{A}$ is the infinitesimal generator associated with an analytic semigroup containing linear operators.

8.2.3 DYNAMICAL BUCKLING

Dynamic hyperbolic equation may be employed to simulate the kinetic buckling of a pivoted extensible beam that is either stretched or compressed by an axial force in a Hilbert space

$$
\frac{\partial^2 w}{\partial \tau^2} + \frac{\partial^4 w}{\partial y^4} - \left(\gamma + \eta \int_0^{\mathscr{L}} \left|\frac{\partial w}{\partial \tau}(\zeta,\tau)\right|^2 d\zeta\right)\frac{\partial^2 w}{\partial y^2} + g\left(\frac{\partial w}{\partial \tau}\right) = 0 \tag{8.2}
$$

where $\gamma, \eta, \mathscr{L} > 0$, $w(\tau,y)$ represents the deviation of the beam's point y during the moment τ, g constitutes a non-decreasing numerical function, and $\mathscr{L}$ indicates the measured length of the beam's deflection.

Equation (8.2) possesses a $\mathbb{R}^n$ equivalent and is able to be used as part of a wider mathematical framework

$$
w'' + \mathscr{A}^2 w + N(\|\mathscr{A}^{1/2}w\|_G^2)\mathscr{A}w + h(w') = 0, \tag{8.3}
$$

where G, N, h are true functions and $\mathscr{A}$ constitutes a linear operator in the case of Hilbert space. The Equations (8.2) and (8.3) were explored by Patcheu and Matos and Pereira, respectively [32] and [33]. These equations represent variations on the second-order damped nonlinear differential model in an amorphous space shown below:

$$
\begin{aligned}
w'' + \mathscr{A}w + \mathscr{B}w' &= f(\tau,w), \\
w(0) = w_0, \ w'(0) &= w_1,
\end{aligned}
$$

where $\mathscr{A}$ and $\mathscr{B}$ are linear operators.

8.2.4 ABSTRACT DIFFERENTIAL EQUATION

8.2.4.1 Abstract First-Order Differential Equations (PDEs as ODEs)

Consider the heat equation

$$
\frac{\partial u(t,x)}{\partial t} = \frac{\partial^2 u(t,x)}{\partial x^2}, \ t \geq 0, \tag{8.4}
$$

in the interval $0 \leq x \leq \pi$, with the initial condition

$$u(0,x) = \phi(x), \ 0 \leq x \leq \pi, \tag{8.5}$$

and the boundary conditions

$$u(t,0) = u(t,\pi) = 0, \ t \geq 0. \tag{8.6}$$

We assume that $u(t,x)$ is the classical solution of the initial boundary value problem, means that $u(t,x)$ is continuous in $[0,\infty) \times [0,\pi]$, and satisfies the initial and boundary conditions, $u_t(t,x)$ exists and is continuous in the same region, $u_x(t,x)$ and $u_{xx}(t,x)$ exist in $[0,\infty) \times (0,\pi)$ and $u(t,x)$ satisfies (8.4). Note that, since $u_{xx} = u_t$, u_{xx} is continuous $[0,\infty) \times [0,\pi]$, the same is true of u_x.

Obviously, $\phi(x)$ itself must be continuous in $[0,\pi]$ and must satisfy the boundary conditions (8.6). We can write this initial boundary value problem as a pure initial value problem for an ordinary differential equation in a Banach space as follows: Let $E = C_0[0,\pi]$ be the space of all continuous function $u(\cdot)$ in $0 \leq x \leq \pi$ with $u(0) = u(\pi) = 0$ equipped with the supremum norm. Define an operator A in E by

$$Au(x) = u''(x) \tag{8.7}$$

with domain $D(A)$ consisting of all functions $u(\cdot) \in C_0[0,\pi]$ twice continuously differentiable in $0 < u < \pi$ with u', u'' continuous in $0 \leq x \leq \pi$ and $u(0) = u(\pi) = 0$. Let $u(t,x)$ be a classical solution of Equations (8.4)–(8.6). Define a function $t \to u(t) \in E$ by $u(t)(x) = u(t,x), 0 \leq x \leq \pi, t \geq 0$. Clearly, $u(t) \in D(A)$, for all t. Moreover, if $t \to y(t) \in E$ is defined by $y(t)(x) = u_t(t,x)$, using the uniform continuity of $u_t(t,x)$ on compact subsets of $[0,\infty) \times [0,\pi]$, we show that, for $t \geq 0$,

$$\left\| \frac{u(t+h) - u(t)}{h} - y(t) \right\| \to 0, \ \text{as } h \to 0. \tag{8.8}$$

It turns out that $u(t)$ is continuously differentiable in the sense of norm of E

$$u'(t) = Au(t), \ t \leq 0, \ u(0) = \phi, \tag{8.9}$$

where ϕ is the initial function in Equation (8.6).

8.2.4.1.1 Solution Procedure

Consider the linear inhomogeneous initial value problem

$$\left. \begin{array}{rcl} \dfrac{dx(t)}{dt} &=& Ax(t) + f(t) \\ x(0) &=& x_0 \end{array} \right\} \tag{8.10}$$

Let $S(t)$ be a C_0- semigroup generated by the infinitesimal generator A and x be a solution of Equation (8.10). Then, the X valued function $\xi(s) = S(t-s)x(s)$ is differentiable on $0 < s < t$ and

$$\begin{array}{rcl} \dfrac{d\xi(s)}{ds} &=& -\dfrac{dS(t-s)}{ds}x(s) + S(t-s)\dfrac{dx(s)}{ds} \\[2mm] &=& -AS(t-s)x(s) + S(t-s)Ax(s) + S(t-s)f(s) \\[2mm] &=& S(t-s)f(s). \end{array}$$

If $f \in L^1(0,T;X)$, then $S(t-s)f(s)$ is integrable, and integrating from 0 to t yields

$$\begin{array}{rcl} \displaystyle\int_0^t \dfrac{d\xi(s)}{ds} &=& \displaystyle\int_0^t S(t-s)f(s,x(s))ds \\[3mm] \xi(t) - \xi(0) &=& \displaystyle\int_0^t S(t-s)f(s,x(s))ds \\[3mm] x(t) &=& S(t)(x_0) + \displaystyle\int_0^t S(t-s)f(s,x(s))ds \end{array} \tag{8.11}$$

Thus, the function $x(t)$ given in Equation (8.11) is called the mild solution of the initial value problem (8.10) on $0 \leq t \leq T$.

The above solution procedure is extended to the nonlinear differential and integrodifferential problems.

8.2.4.2 Abstract Second-Order Equations

Several problems involving physical processes, such as the vibration of pivoted bars and the transverse motion of an extendable beam, include second-order nonlinear differential and integrodifferential equations. Studying the controllability issue associated with such systems in Banach spaces is therefore extremely significant. Instead of converting the second-order abstract differential equations to lower order systems, it is frequently beneficial to handle them directly. For instance, a problem as simple as the growth of a solution to the growth equation does not benefit greatly from reduction to first order

$$z''(\tau) = (\mathscr{A} + cI)z(\tau), \tag{8.12}$$

in terms of the growth of solution of

$$z''(\tau) = \mathscr{A}z(\tau).$$

Direct study of the subsequent order equation yields simpler and more comprehensive theories for other issues like singular perturbation. Finally, the integral formulation associated with $z''(\tau) = \mathscr{A}z(\tau)$, $z(0) = z_0$, $z'(0) = y_0$ has proved useful in giving motivation to the study of second-order differential equations in Banach spaces. Referrals for a mathematical formulation and physical justification are available [21, 22].

8.2.5 METHODS

8.2.5.1 Fixed Point Method

The fixed point approach is the most effective way for studying the controllability of nonlinear integrodifferential systems of all the methods. The fixed point approach is used to prove integrodifferential equations' existence theorems and to investigate the controllability problem for integrodifferential systems. Due to its significance, a number of academics have investigated the challenges posed by evolution equations using various types of fixed point theorems [39, 40]. In this method, the problem is turned into a fixed point problem in a function space using an appropriate nonlinear operator. Furthermore, controllability criteria in the Banach spaces of continuous functions can be obtained using fixed point theorems. The main focus of this chapter is on using the Banach contraction principle to prove controllability conclusions for second-order integrodifferential systems in Banach spaces.

8.2.5.2 Strongly Continuous Cosine Families

Most notably, the idea of extremely continuous cosine families of limited linear operators is used to resolve boundary value and beginning problems for second-order PDEs. The relationship between abstract linear second-order differential equations and strongly continuous cosine families of bounded linear operators is analogous to the relationship between abstract linear first-order differential equations and strongly continuous semi-groups of bounded linear operators. The extremely continuous cosine collection of bounded linear operators along with infinitesimal generator $\mathscr{A}$ follows any well-posed second-order differential equation of the kind $z'' = \mathscr{A}w$, and vice-versa. The most essential and thorough research on cosine families has been done by Travis and Webb [28, 29, 34].

Throughout this chapter, one can study the controllability of nonlinear structures characterized by second-order neutral integrodifferntial equations employing the cosine function theory and the contraction mapping notion, as prompted by Refs. [35–38].

8.3   PROBLEM FORMULATION

Lets consider the subsequent structure

$$\frac{d}{dt}\left[z'(t)+h(t,z(t),z'(t))\right] = \mathscr{A}z(t)+\mathscr{B}v(t)+f\Big(t,z(t),z'(t)\Big)$$

$$+\int_0^t g\Big(t,s,z(s),z'(s)\Big)ds, \ t\in\mathbb{I}, \tag{8.13}$$

$$z(0) = z_0, \quad z'(0)=x_0, \tag{8.14}$$

wherein

- the state variable $z(\cdot)$ takes values in a Banach space $\mathbb{X}$;
- $\mathscr{A}$ represents the infinitesimal generator that produces a substantially continuous cosine function of bounded linear operators over a Banach space $\mathbb{X}$ and $\mathbb{I}=[0,a]$;
- The control function $v(\cdot)$ is supplied in $\mathbb{L}^2(\mathbb{I},\mathbb{V})$, a Banach space containing admissible control functions, with $\mathbb{V}$ serving as the Banach space;
- $\mathscr{B}$ constitutes a bounded linear operator spanning $\mathbb{V}$ to $\mathbb{X}$;
- $h(\cdot),f(\cdot):\mathbb{I}\times\mathbb{X}\times\mathbb{X}\to\mathbb{X}$, $g(\cdot):\mathbb{I}\times\mathbb{I}\times\mathbb{X}^2\to\mathbb{X}$ are given appropriate functions.

8.3.1   PRELIMINARIES

In what follows, we put $t_0=0$, $t_{n+1}=a$ and we denote, by $\mathbb{C}$, the space formed by the functions $z:\mathbb{I}\to\mathbb{X}$ such that $z(\cdot)$ is continuous. It is clear that $\mathbb{C}$, endowed with the norm $\|z\|_{\mathbb{C}}:=\sup_{t\in\mathbb{I}}\|z(t)\|$, is a Banach space. Similarly, $\mathbb{C}'$ will be the space of the functions $z(\cdot)\in\mathbb{C}$ such that $z(\cdot)$ is continuously differentiable on $\mathbb{I}\backslash t_j$, $j=1,2,\ldots,m$ and the derivatives $w_R'(t)=\lim\limits_{s\to 0}\dfrac{w(t+s)-w(t^+)}{s}$, $w_L'(t)=\lim\limits_{s\to 0}\dfrac{w(t+s)-w(t^-)}{s}$ are continuous on $[0,a]$ and $[0,a]$ respectively. It is easy to see that $\mathbb{C}'$, provided with the norm

$$\|w\|_{\mathbb{C}'}:=\|w\|_{\mathbb{C}}+\|w'\|_{\mathbb{C}},$$

is a Banach space.

The operator-valued function $\mathscr{H}(t)=\begin{bmatrix}\mathscr{C}(t) & \mathscr{S}(t)\\ \mathscr{A}\mathscr{S}(t) & \mathscr{C}(t)\end{bmatrix}$ is a strongly continuous group of linear operators on the space $\mathscr{E}\times\mathbb{X}$ generated by the operator $\mathscr{A}=\begin{bmatrix}0 & \mathbb{I}\\ \mathscr{A} & 0\end{bmatrix}$ defined on $D(\mathbb{A})\times\mathscr{E}$. From this, it follows that $\mathscr{A}\mathscr{S}(t):\mathscr{E}\to\mathbb{X}$ is a bounded linear operator and that $\mathscr{A}\mathscr{S}(t)z\to 0$ as $t\to 0$ for each $z\in\mathscr{E}$. Furthermore, if $z:[0,\infty]\to\mathbb{X}$ is locally integrable, then $y(t):=\int_0^t\mathscr{S}(t-s)z(s)ds$ defines an $\mathscr{E}$- valued continuous function, which is a consequence of the fact that $\int_0^t\mathscr{H}(t-s)\begin{bmatrix}0\\ z(s)\end{bmatrix}ds=\begin{bmatrix}\int_0^t\mathscr{S}(t-s)z(s)ds\\ \int_0^t\mathscr{C}(t-s)z(s)ds\end{bmatrix}$ defines an $\mathscr{E}\times\mathbb{X}-$ valued continuous function.

Definition 8.1 *[28] A one parameter family $\{\mathscr{C}(t),\ t\in\mathscr{R}\}$ of bounded linear operators in the Banach space $\mathbb{X}$ is called a strongly continuous cosine family if and only if*

- $\mathscr{C}(s+t)+\mathscr{C}(s-t)=2\mathscr{C}(s)\mathscr{C}(t),\ \ for\ all\ \ s,t\in\mathscr{R};$
- $\mathscr{C}(0)=\mathbb{I};$
- $\mathscr{C}(t)z$ *is continuous in t on $\mathscr{R}$, for each fixed $z\in\mathbb{X}$.*

Define the associated sine family $\{\mathscr{S}(t),t\in\mathscr{R}\}$ by

$$\mathscr{S}(t)z:=\int_0^t\mathscr{C}(s)zds,\quad z\in\mathbb{X},\quad t\in\mathscr{R}.$$

The infinitesimal generator of a strongly continuous cosine family $\{\mathscr{C}(t),\ t \in \mathscr{R}\}$ is the operator $\mathscr{A} : \mathbb{X} \to \mathbb{X}$, defined by

$$\mathscr{A}z = \frac{d^2}{dt^2}\mathscr{C}(t)z|_{t=0}, \qquad z \in D(\mathscr{A}),$$

where $D(\mathscr{A}) := -z \in \mathbb{X}$: $\mathscr{C}(t)z$ is twice continuously differentiable in $t\ ˝$.
Define $\mathscr{E} := -z \in \mathbb{X}$: $\mathscr{C}(t)z$ is continuously differentiable in $t\ ˝$. We assume

(**B1**) $\mathscr{A}$ is the infinitesimal generator of a strongly continuous cosine family $\{\mathscr{C}(t),\ t \in \mathscr{R}\}$ of bounded linear operators in the Banach space $\mathbb{X}$.

To establish our main theorem, we need the following lemmas:

Lemma 8.1 *[28] Let* (**B1**) *hold. Then,*

- *there exist constants $M \geq 1$ and $\omega \geq 0$ such that $\|\mathscr{C}(t)\| \leq Me^{\omega|t|}$ and*

$$\|\mathscr{S}(t) - \mathscr{S}(t^*)\| \leq M\left|\int_t^{t^*} e^{\omega|s|}ds\right|,\ \text{for}\ t,\ t^* \in \mathscr{R};$$

- $\mathscr{S}(t)\mathbb{X} \subset \mathscr{E}$ *and* $\mathscr{S}(t)\mathscr{E} \subset D(\mathscr{A})$,   *for $t \in \mathscr{R}$;*

- $\dfrac{d}{dt}\mathscr{C}(t)z = \mathscr{A}\mathscr{S}(t)z$,   *for* $z \in \mathscr{E}$ *and* $t \in \mathscr{R}$;

- $\dfrac{d^2}{dt^2}\mathscr{C}(t)x = \mathscr{A}\mathscr{C}(t)z$,   *for* $z \in D(\mathscr{A})$ *and* $t \in \mathscr{R}$.

Lemma 8.2 *[28] Let* (**B₁**) *hold and $v : \mathscr{R} \to \mathbb{X}$ be such that u is continuous and let $p(t) = \int_0^t \mathscr{S}(t-s)u(s)ds$. Then, p is twice continuously differentiable and, for $t \in \mathscr{R}$, $p(t) \in D(\mathscr{A})$, $p'(t) = \int_0^t \mathscr{C}(t-s)u(s)ds$ and $p''(t) = \int_0^t \mathscr{C}(t-s)u'(s)ds + \mathscr{C}(t)u(0) = \mathscr{A}p(t) + u(t)$.*

First, take into account the neutral system

$$\frac{d}{dt}[z'(t) + h(t,z(t))] = \mathscr{A}z(t) + \mathscr{B}v(t) + f(t,z(t))$$

$$+ \int_0^t g(t,s,z(s))ds,\ t \in \mathbb{I},\ t \neq t_j, \tag{8.15}$$

$$z(0) = z_0,\ z'(0) = x_0. \tag{8.16}$$

If $z(\cdot)$ is a solution of Equations (8.15) and (8.16) and $h(t,z(t))$ is sufficiently smooth, then (see Ref. [45]) we get

$$z(t) = \mathscr{C}(t)z_0 + \mathscr{S}(t)[x_0 + h(0,z(0))] - \int_0^t \mathscr{C}(t-s)h(s,z(s))ds + \int_0^t \mathscr{S}(t-s)\mathscr{B}v(s)ds$$

$$+ \int_0^t \mathscr{S}(t-s)f(s,z(s))ds + \int_0^t \mathscr{S}(t-s)\int_0^s g(s,\tau,z(\tau))d\tau ds.$$

This observation inspires the prospect of a mild solution spanning the system (8.15) and (8.16).

Definition 8.2 *Since the integral equation for a function $z \in \mathbb{C}(\mathbb{I},\mathbb{X})$ is satisfied, then the system (8.15) and (8.16) is deemed to have a mild solution.*

$$z(t) = \mathscr{C}(t)z_0 + \mathscr{S}(t)[x_0 + h(0,z(0))] - \int_0^t \mathscr{C}(t-s)h(s,z(s))ds + \int_0^t \mathscr{S}(t-s)\mathscr{B}v(s)ds$$

$$+ \int_0^t \mathscr{S}(t-s)f(s,z(s))ds + \int_0^t \mathscr{S}(t-s)\int_0^s g(s,\tau,z(\tau))d\tau ds,\ t \in \mathbb{I}$$

is verified.

If the problem (8.15) and (8.16) possesses a mild solution, $z(\cdot)$, then, according to Lemma 8.2 and the characteristics of a second-order differential equation, we get

$$
\begin{aligned}
z'(t) \;=\; & \mathscr{A}\mathscr{S}(t)z_0 + \mathscr{C}(t)[x_0 + h(0,z(0))] - h(t,z(t)) - \int_0^t \mathscr{A}\mathscr{S}(t-s)h(s,z(s))ds \\
& + \int_0^t \mathscr{C}(t-s)\mathscr{B}v(s)ds + \int_0^t \mathscr{C}(t-s)f(s,z(s))ds + \int_0^t \mathscr{C}(t-s)\int_0^s g(s,\tau,z(\tau))d\tau ds, \; t \in \mathbb{I}.
\end{aligned}
$$

Let

$$
\mathscr{B}_r := \{z \in \mathbb{X} : \|z\| \leq r\} \text{ for some } r \geq 1.
$$

Let's provide the following hypotheses in order to acquire the controllability results:

(A1) $\mathscr{A}$ is the infinitesimal generator of a strongly continuous cosine family $\{\mathscr{C}(t),\, t \in \mathscr{R}\}$ of bounded linear operators in the Banach space $\mathbb{X}$. There exist constants $N_1 \geq 1$ and $N_2 \geq 0$ such that $\|\mathscr{C}(t)\| \leq N_1$ and $\|\mathscr{S}(t)\| \leq N_2$, for every $t \in [0,a]$. Furthermore we take $N_3 = \sup_{0 \leq t \leq a} \|\mathscr{A}\mathscr{S}(t)\|$.

(A2) The linear operator $\mathscr{G}_1 : \mathbb{L}^2(\mathbb{I},\mathbb{V}) \to \mathbb{X}$, defined by

$$
\mathscr{G}_1 u = \int_0^a \mathscr{S}(a-s)\mathscr{B}v(s)ds,
$$

has an inverse $\mathscr{G}_1^{-1}$ which takes values in $\mathbb{L}^2(\mathbb{I},\mathbb{V})/ker\mathscr{G}_1$, and there exists a positive constant $\mathscr{K}_1$ such that $\|\mathscr{B}\mathscr{G}_1^{-1}\| \leq \mathscr{K}_1$.

(A3) The linear operator $\mathscr{G}_2 : \mathbb{L}^2(\mathbb{I},\mathbb{V}) \to \mathbb{X}$, defined by

$$
\mathscr{G}_2 v = \int_0^a \mathscr{C}(a-s)\mathscr{B}v(s)ds,
$$

has an inverse $\mathscr{G}_2^{-1}$ which takes values in $\mathbb{L}^2(\mathbb{I},\mathbb{V})/ker\mathscr{G}_2$, and there exists a positive constant $\widehat{\mathscr{K}_1}$ such that $\|\mathscr{B}\mathscr{G}_2^{-1}\| \leq \widehat{\mathscr{K}_1}$.

(A4) $\mathscr{G}_1\mathscr{G}_2^{-1}z = \mathscr{G}_2\mathscr{G}_1^{-1}z = 0$, for every $z \in \mathbb{X}$.

(A5) The function $h : \mathbb{I} \times \mathbb{X} \to \mathbb{X}$ satisfies the following conditions:

   (i) The function $h(t,.) : \mathbb{X} \to \mathbb{X}$ is continuous a.e., $t \in \mathbb{I}$.

  (ii) The function $h(.,z) : \mathbb{I} \to \mathbb{X}$ is strongly measurable, for each $z \in \mathbb{X}$.

 (iii) There exist positive constants $\mathscr{K}_h > 0$, $\widehat{\mathscr{K}_h} > 0$ such that

$$
\|h(t,x(t)) - h(s,y(t))\| \leq \mathscr{K}_h[|t-s| + \|x-y\|], \text{ for } t,s \in \mathbb{I}, \; x,\, y \in \mathbb{X}
$$

   and $\widehat{\mathscr{K}_h} = \max_{t \in \mathbb{I}} \|h(t,0)\|$.

(A6) The function $f : \mathbb{I} \times \mathbb{X} \to \mathbb{X}$ satisfies the following conditions:

   (i) For each $t \in \mathbb{I}$, the function $f(t,.) : \mathbb{X} \to \mathbb{X}$ is continuous, and for each $z \in \mathbb{X}$, the function $f(.,z) : \mathbb{I} \to \mathbb{X}$ is strongly measurable.

  (ii) There exist positive constants $\mathscr{K}_f > 0$, $\widehat{\mathscr{K}_f} > 0$ such that

$$
\|f(t,z_1) - f(t,z_2)\| \leq \mathscr{K}_f\|z_1 - z_2\|, \text{ for any } t \in \mathbb{I}, \; z_i \in \mathbb{X}, i = 1,2,
$$

   and $\widehat{\mathscr{K}_f} = \max_{t \in \mathbb{I}} \|f(t,0)\|$.

(A7) The function $g(t,s,.) : \mathbb{X} \to \mathbb{X}$ is continuous, and for each $x \in X$, the function $g(.,.,z) : \mathbb{I} \times \mathbb{I} \to \mathbb{X}$ is strongly measurable. There exist constants $\mathscr{K}_g > 0$, $\widehat{\mathscr{K}_g} > 0$ such that

$$
\|g(t,s,y) - h(t,s,z)\| \leq \mathscr{K}_g\|y - z\|, \text{ for any } t,s \in I, \; x,y \in X,
$$

and $\widehat{\mathscr{K}_g} = \max_{t,s \in \mathbb{I}} \|g(t,s,0)\|$.

(A8) There exist constants $r > 0$, $\widehat{r} > 0$ such that

$$N_1\|z_0\| + N_2[\|z_0\| + \widehat{\mathscr{K}_h}] + aN_1[\mathscr{K}_g r + \widehat{\mathscr{K}_h}] + aN_2\mathscr{K}_0$$
$$+aN_2[\mathscr{K}_f \rho + \widehat{\mathscr{K}_f}] + aN_2[\mathscr{K}_g r + \widehat{\mathscr{K}_g}] \;\leq\; r,$$

$$N_3\|z_0\| + N_1[\|x_0\| + \widehat{\mathscr{K}_h}] + (1 + aN_3)[\mathscr{K}_h r + \widehat{\mathscr{K}_h}] + aN_1\mathscr{K}_0$$
$$+aN_1[\mathscr{K}_f r + \widehat{\mathscr{K}_f}] + aN_1[\mathscr{K}_g r + \widehat{\mathscr{K}_g}] \;\leq\; \widehat{r}.$$

For convenience, let us put

$$\gamma \;=\; (1 + aN_2\mathscr{K}_1)\Big[aN_1\mathscr{K}_h + bN_2\mathscr{K}_f + a^2 N_2\mathscr{K}_g\Big]$$
$$+aN_2\widehat{\mathscr{K}_1}\Big(\mathscr{K}_h + aN_3\mathscr{K}_g + aN_1\mathscr{K}_f + a^2 N_1\mathscr{K}_g\Big)$$

and

$$\widehat{\gamma} \;=\; aN_1\mathscr{K}_1\Big(bN_1\mathscr{K}_h + aN_2\mathscr{K}_f + a^2 N_2\mathscr{K}_g\Big)$$
$$+[1 + aN_1\widehat{\mathscr{K}_1}]\Big(\mathscr{K}_g + aN_3\mathscr{K}_g + aN_1\mathscr{K}_f + a^2 N_1\mathscr{K}_g\Big).$$

Definition 8.3 *(Banach Fixed Point Theorem [12]) Let $\mathbb{W}$ be a closed subset of the Banach space $\mathbb{X}$, $\mathscr{T}$ a mapping from $\mathbb{W}$ to $\mathbb{W}$, $m \in \mathbb{N}$, and $\alpha < 1$. Suppose that $\mathscr{T}$ satisfies*

$$\|T^m(x_1) - T^m(x_2)\| \leq \alpha\|x_1 - x_2\|,$$

for all $x_1, x_2 \in \mathbb{W}$. Then, there exists a unique $x^* \in \mathbb{W}$ such that $T(x^*) = x^*$. The point x^* is called the fixed point of $\mathscr{T}$.

Definition 8.4 *The system (8.15) and (8.16) is said to be controllable on the interval $\mathbb{I} = [0, a]$, for every $z_0, z_a \in D(\mathscr{A})$ and $x_0, x_a \in \mathscr{E}$, if there exists a control $v \in \mathbb{L}^2(\mathbb{I}, \mathbb{V})$ such that the solution $z(\cdot)$ of Equation (8.15) and (8.16) satisfies $z(a) = z_a$, $z'(a) = x_a$.*

8.4 CONTROLLABILITY RESULTS

Theorem 8.1 *If the assumptions (A1)–(A8) are satisfied and if $0 \leq \gamma, \widehat{\gamma} < 1$, then the system Equations (8.15) and (8.16) is controllable on I.*

Proof. Using (A2) and (A3) for an arbitrary function $x(\cdot)$, we define the control

$$v_z(t) \;=\; \mathscr{G}_1^{-1}\Big[x_a - \mathscr{C}(a)z_0 - \mathscr{S}(a)[x_0 + h(0, z(0))] + \int_0^a \mathscr{C}(a - s)h(s, x(s))ds$$
$$- \int_0^a \mathscr{S}(a - s)f(s, x(s))ds - \int_0^a \mathscr{S}(a - s)\int_0^s g(s, \tau, x(\tau))d\tau ds\Big](t)$$
$$+ \mathscr{G}_2^{-1}\Big[z_a - \mathscr{A}\mathscr{S}(a)z_0 - \mathscr{C}(a)[x_0 + h(0, z(0))] + h(a, z(a)) + \int_0^a \mathscr{A}\mathscr{S}(a - s)h(s, z(s))ds$$
$$- \int_0^a \mathscr{C}(a - s)f(s, x(s))ds - \int_0^a \mathscr{C}(a - s)\int_0^s g(s, \tau, z(\tau))d\tau ds\Big](t).$$

Now we have to show that, when using this control, the operator

$$\Psi : \mathbb{C}(\mathbb{I}, \mathbb{B}_r) \to \mathbb{C}(\mathbb{I}, \mathbb{B}_r),$$

defined by

$$
\begin{aligned}
(\Psi z)(t) \;=\; & \mathscr{C}(t)z_0 + \mathscr{S}(t)[z_0 + h(0,z(0))] - \int_0^t \mathscr{C}(t-s)h(s,x(s))ds \\
& + \int_0^t \mathscr{S}(t-s)\Big\{ \mathscr{BG}_1^{-1}\Big[x_a - \mathscr{C}(a)z_0 - \mathscr{S}(a)[z_0 + h(0,z(0))] \\
& + \int_0^a \mathscr{C}(a-s)h(s,x(s))ds - \int_0^a \mathscr{S}(a-s)f(s,x(s))ds \\
& - \int_0^a \mathscr{S}(a-s)\int_0^s g(s,\tau,x(\tau))d\tau ds\Big](s) + \mathscr{BG}_2^{-1}\Big[x_a - \mathscr{A}\mathscr{S}(a)z_0 \\
& - \mathscr{C}(a)[z_0 + h(0,z(0))] + h(a,z(a)) + \int_0^a \mathscr{A}\mathscr{S}(a-s)h(s,z(s))ds \\
& - \int_0^a \mathscr{C}(a-s)f(s,z(s))ds - \int_0^a \mathscr{C}(a-s)\int_0^s g(s,\tau,z(\tau))d\tau ds\Big](s)\Big\}ds \\
& + \int_0^t \mathscr{S}(t-s)f(s,z(s))ds + \int_0^t \mathscr{S}(t-s)\int_0^s g(s,\tau,z(\tau))d\tau ds,
\end{aligned}
$$

has a fixed point $z(\cdot)$, which is a solution of Equations (8.15) and (8.16). Clearly, $z(a) = z_a$, $z'(a) = x_a$, which imply that the system is controllable. Since all the functions involved in the operator are continuous, then Ψ is continuous. Based on our assumptions

$$
\begin{aligned}
\|\mathscr{Q}(\zeta,z)\| \;\le\; & \mathscr{K}_1\Big[\|z_a\| + N_1\|z_0\| + N_2[\|x_0\| + \widehat{\mathscr{K}_h}] + aN_1[\mathscr{K}_h r + \widehat{\mathscr{K}_h}] + aN_2[\mathscr{K}_f r + \widehat{\mathscr{K}_f}] \\
& + a^2 N_2[\mathscr{K}_g r + \widehat{\mathscr{K}_g}]\Big] \\
& + \widehat{\mathscr{K}_1}\Big[\|x_a\| + N_3\|z_0\| + N_1[\|x_0\| + \widehat{\mathscr{K}_h}] + (1 + aN_3)[\mathscr{K}_h r + \widehat{\mathscr{K}_h}] \\
& + aN_1[\mathscr{K}_f r + \widehat{\mathscr{K}_f}] + a^2 N_1[\mathscr{K}_g r + \widehat{\mathscr{K}_g}]\Big] \\
\;=\; & \mathscr{K}_0
\end{aligned}
$$

and

$$
\begin{aligned}
\|\mathscr{Q}(\zeta,z_1) - \mathscr{Q}(\zeta,z_2)\| \;\le\; & \Big\{ \mathscr{K}_1\Big(aN_1\mathscr{K}_h + aN_2\mathscr{K}_f + a^2 N_2\mathscr{K}_g\Big) \\
& + \widehat{\mathscr{K}_1}\Big(\mathscr{K}_h + aN_3\mathscr{K}_h + aN_1\mathscr{K}_f + a^2 N_1\mathscr{K}_g\Big)\Big\}\|z_1 - z_2\|.
\end{aligned}
$$

First, we show that Ψ maps $\mathbb{C}(\mathbb{I}, \mathbb{B}_r)$ into itself. Now

$$
\begin{aligned}
\|(\Psi z)(t)\| \;\le\; & N_1\|z_0\| + N_2[\|z_0\| + \widehat{\mathscr{K}_h}] + aN_1[\mathscr{K}_h r + \widehat{\mathscr{K}_h}] + aN_2\mathscr{K}_0 + aN_2[\mathscr{K}_f r + \widehat{\mathscr{K}_f}] \\
& + aN_2[\mathscr{K}_g r + \widehat{\mathscr{K}_h}]
\end{aligned}
$$

and

$$
\begin{aligned}
\|(\Psi z)'(t)\| \;\le\; & N_3\|z_0\| + N_1[\|z_0\| + \widehat{\mathscr{K}_h}] + (1 + aN_3)[\mathscr{K}_h r + \widehat{\mathscr{K}_h}] + aN_1\mathscr{K}_0 \\
& + aN_1[\mathscr{K}_f r + \widehat{\mathscr{K}_f}] + aN_1[\mathscr{K}_g r + \widehat{\mathscr{K}_h}].
\end{aligned}
$$

From the assumption $(A8)$, $\|(\Psi z)(t)\| \le r$ and $\|(\Phi z)'(t)\| \le \widehat{r}$. Therefore, Ψ maps $\mathbb{C}(\mathbb{I}, \mathbb{B}_r)$ into itself. Moreover, if $z_1, z_2 \in \mathbb{C}(\mathbb{I}, \mathbb{B}_r)$, then

$$
\begin{aligned}
\|(\Psi z_1)(t) - (\Psi z_2)(t)\| &\leq \left\| \int_0^t \mathscr{C}(t-s)[h(s,z_1(s)) - h(s,z_2(s))]ds \right\| \\
&\quad + \left\| \int_0^t \mathscr{S}(t-s)[\mathscr{Q}(\zeta,z_1) - \mathscr{Q}(\zeta,z_2)]d\eta \right\| \\
&\quad + \left\| \int_0^t \mathscr{S}(t-s)[f(s,x(s)) - f(s,y(s))]ds \right\| \\
&\quad + \left\| \int_0^t \mathscr{S}(t-s)\left[\int_0^s h(s,\tau,z_1(\tau))d\tau - \int_0^s h(s,\tau,z_2(\tau))d\tau \right]ds \right\| \\
&\leq \left\{ (1 + aN_2\mathscr{K}_1)\left[aN_1\mathscr{K}_h + aN_2\mathscr{K}_f + b^2 N_2\mathscr{K}_h \right. \right. \\
&\quad \left. + N_1 d_i^1 + N_2 d_i^2 \right] + aN_2\widehat{\mathscr{K}_1}\left(\mathscr{K}_h + aN_3\mathscr{K}_h \right. \\
&\quad \left. \left. + aN_1\mathscr{K}_f + a^2 N_1\mathscr{K}_h \right) \right\} \|z_1 - z_2\| \\
&= \gamma \|z_1 - z_2\|_{\mathscr{P}\mathscr{C}}. \tag{8.17}
\end{aligned}
$$

Also,

$$
\begin{aligned}
\|(\Psi z_1)'(t) - (\Psi z_2)'(t)\| &\leq \left\{ aN_1\mathscr{K}_1\left(aN_1\mathscr{K}_h + aN_2\mathscr{K}_f + a^2 N_2\mathscr{K}_g + N_1 d_i^1 + N_2 d_i^2 \right) \right. \\
&\quad \left. + [1 + aN_1\widehat{\mathscr{K}_1}]\left(\mathscr{K}_h + aN_3\mathscr{K}_h + aN_1\mathscr{K}_f + a^2 N_1\mathscr{K}_g \right) \right\} \|z_1 - z_2\| \\
&= \widehat{\gamma} \|z_1 - z_2\|_{\mathscr{P}\mathscr{C}}. \tag{8.18}
\end{aligned}
$$

Since $\gamma < 1$ and $\widehat{\gamma} < 1$, Equations (8.17) and (8.18) show that the operator Φ is a contraction on $\mathbb{C}(\mathbb{I}, \mathbb{B}_r)$ and so, by Banach's fixed point theorem, there exists a unique fixed point $z \in \mathbb{C}(\mathbb{I}, \mathbb{B}_r)$ such that $z(a) = (\Psi z)(a) = z_a$, $z'(a) = (\Psi z)'(a) = x_a$ implying that the system (8.15) and (8.16) is controllable on $\mathbb{I}$. Thus, Theorem 8.1 is proved. $\qquad \square$

Next, we discuss the controllability of the system (8.13) and (8.14). For this, we impose the following conditions:

(A9) The function $h : \mathbb{I} \times \mathbb{X} \times \mathbb{X} \to \mathbb{X}$ satisfies the following conditions:
(i)The function $h(t,.,.) : \mathbb{X} \times \mathbb{X} \to \mathbb{X}$ is continuous a.e., $t \in \mathbb{I}$.
(ii)The function $h(.,y,z) : \mathbb{I} \to \mathbb{X}$ is strongly measurable, for each $y, z \in \mathbb{X}$.
(iii)There exist positive constants $\mathscr{K}_h^1 > 0$, $\widetilde{\mathscr{K}_h^1} > 0$ such that

$$
\|h(t,y_1,z_1) - h(s,y_2,z_2)\| \leq \mathscr{K}_g^1 [|t-s| + \|y_1 - y_2\| + \|z_1 - z_2\|],
$$

for $t, s \in \mathbb{I}$, $y_i, z_i, \in X, i = 1, 2$, and

$$
\widetilde{\mathscr{K}_h^1} = \max_{t \in \mathbb{I}} \|h(t,0,0)\|.
$$

(A10) The function $f : \mathbb{I} \times \mathbb{X} \times \mathbb{X} \to \mathbb{X}$ satisfies the following conditions:
(i) For each $t \in \mathbb{I}$, the function $f(t,.,.) : \mathbb{X} \times \mathbb{X} \to \mathbb{X}$ is continuous, and for each $y, z \in \mathbb{X}$, the function $f(.,y,z) : \mathbb{I} \to \mathbb{X}$ is strongly measurable.
(ii) There exist positive constants $\mathscr{K}_f^1 > 0$, $\widetilde{\mathscr{K}_f^1} > 0$ such that

$$
\|f(t,y_1,z_1) - f(t,y_2,z_2)\| \leq \mathscr{K}_f^1 [\|y_1 - y_2\| + \|z_1 - z_2\|],
$$

for any $t \in \mathbb{I}$, $y_i, z_i \in \mathbb{X}, i = 1, 2$, and

$$\widetilde{K_f^1} = \max_{t \in \mathbb{I}} \|f(t,0,0)\|.$$

(A11) The function $g(t,s,.,.) : \mathbb{X} \times \mathbb{X} \to \mathbb{X}$ is continuous, and for each $y,z \in \mathbb{X}$, the function $g(.,.,y,z) : \mathbb{I} \times \mathbb{I} \to \mathbb{X}$ is strongly measurable. There exist constants $\mathscr{K}_g^1 > 0$, $\widetilde{\mathscr{K}_g^1} > 0$ such that

$$\|h(t,s,y_1,z_1) - h(t,s,y_2,z_2)\| \leq K_h^1[\|y_1 - y_2\| + \|z_1 - z_2\|],$$

for any $t,s \in \mathbb{I}$, $y_i,z_i \in \mathbb{X}$, $i = 1,2,$

and

$$\widetilde{\mathscr{K}_g^1} = \max_{t,s \in I} \|h(t,s,0,0)\|.$$

We consider the following concept of a mild solution.

Definition 8.5 *A function $z \in \mathscr{C}'(\mathbb{I}, \mathbb{B}_r)$ is said to be a mild solution of the problem (8.13) and (8.14), if it satisfies the following equation*

$$
\begin{aligned}
z(t) \;=\;& \mathscr{C}(t)z_0 + \mathscr{S}(t)[x_0 + h(0,z(0),z'(0))] - \int_0^t \mathscr{C}(t-s)h(s,z(s),z'(s))ds \\
&+ \int_0^t \mathscr{S}(t-s)\mathscr{B}v_z(s)ds + \int_0^t \mathscr{S}(t-s)f(s,z(s),z'(s))ds \\
&+ \int_0^t \mathscr{S}(t-s)\int_0^s g(s,\tau,z(\tau),z'(\tau))d\tau ds, \quad t \in \mathbb{I}.
\end{aligned}
\tag{8.19}
$$

Using this definition, it is worthwhile to point out that if $z(\cdot)$ is a mild solution of the problem (8.13) and (8.14), then, by Lemma 8.2, we have

$$
\begin{aligned}
z'(t) \;=\;& \mathscr{A}\mathscr{S}(t)x_0 + \mathscr{C}(t)[x_0 + h(0,z(0),z'(0))] - h(t,z(t),z'(t)) - \int_0^t \mathscr{A}\mathscr{S}(t-s) \\
&\times h(s,z(s),z'(s))ds + \int_0^t \mathscr{C}(t-s)\mathscr{B}v_z(s)ds + \int_0^t \mathscr{C}(t-s)f(s,z(s),z'(s))ds \\
&+ \int_0^t \mathscr{C}(t-s)\int_0^s g(s,\tau,z(\tau),z\prime(\tau))d\tau ds \quad t \in \mathbb{I}.
\end{aligned}
\tag{8.20}
$$

Proceeding as before and using the assumptions $(A9) - (A11)$ and Equations (8.19) and (8.20), we can establish the controllability results for the system (8.13) and (8.14).

8.5 NON-LOCAL CONTROLLABILITY

This subsection discusses the non-local controllability associated with the nonlinear second-order integrodifferential system (8.13) and (8.14). The non-local condition, an extrapolation of the traditional initial condition, was motivated by physical difficulties. Byszewski was the first to study non-local conditions in Banach spaces, citing [41] as his source. Following that, he looked at a similar problem for several types of evolution equations in Banach spaces [42, 43]. In Banach spaces, Balachandran and Park [44] studied second-order nonlinear differential equations with non-local conditions. For the boundary controllability of differential equations with non-local conditions, Han and Park [45] developed necessary requirements. The non-local condition is defined in this section as

$$z(0) = z_0 + c(z), \quad z'(0) = x_0 + d(z),\tag{8.21}$$

where continuous functions c and d are supplied.

We require the following extra condition in order to demonstrate the controllability result:

(A13) The functions $c,d : \mathbb{C}(\mathbb{I},\mathbb{X}) \to \mathbb{X}$ are continuous, and there exist constants $K_c > 0$, $K_d > 0$ such that

$$\begin{aligned}
\|c(z_1) - c(z_2)\| &\leq K_p\|z_1 - z_2\|_a, \\
\|d(z_1) - d(z_2)\| &\leq K_q\|z_1 - z_2\|_a, \quad \text{for } z_1, z_2 \in \mathscr{PC}(\mathbb{I},\mathbb{X}).
\end{aligned}$$

A function $z \in \mathscr{C}'(\mathbb{I},\mathbb{B}_r)$ is said to be a mild solution of the problem (8.13) and (8.14) with the non-local condition (8.21) if it satisfies the following equation

$$\begin{aligned}
z(t) \;=\; & \mathscr{C}(t)(z_0 + c(z)) + \mathscr{S}(t)[x_0 + d(z) + h(0,z(0),z'(0))] - \int_0^t \mathscr{C}(t-s)h(s,z(s),z'(s))ds \\
& + \int_0^t \mathscr{S}(t-s)\mathscr{B}v_z(s)ds + \int_0^t \mathscr{S}(t-s)f(s,z(s),x'(s))ds \\
& + \int_0^t \mathscr{S}(t-s)\int_0^s g(s,\tau,z(\tau),z'(\tau))d\tau ds, \quad t \in \mathbb{I}.
\end{aligned} \tag{8.22}$$

Definition 8.6 *The system (8.13) and (8.14) with the non-local condition (8.21) is said to be controllable on the interval $\mathbb{I} = [0,a]$, if for every $z_0, z_a \in D(\mathscr{A})$ and $x_0, x_a \in \mathscr{E}$, there exists a control $v \in L^2(\mathbb{I},\mathbb{V})$ such that the solution $x(\cdot)$ of Equations (8.13) and (8.14) with (8.21) satisfies $z(a) = z_a$, $z'(a) = x_a$.*

It should be noted that Theorem 8.1's method can be used to establish the controllability of the system (8.13) and (8.14) with (8.21).

8.6 CONTROLLABILITY OF FINITE DIMENSIONAL SYSTEMS

In this section, second-order neutral integrodifferential systems with impulsive conditions that are controllable in low-dimensional spaces are discussed. Second-order matrix systems are commonly used to model real-world issues. For matrix second-order linear systems to be controllable, Hughes and Skelton [46] supplied the required and sufficient requirements.

Consider the linear system

$$\frac{d^2z(t)}{dt^2} + \mathscr{A}^2 z(t) = \mathscr{B}v(t), \quad z(0) = z_0, \; z'(0) = x_0. \tag{8.23}$$

Many academics have looked into the system (8.23) since it accurately represents the dynamics of many natural events (see Hughes and Skelton [46] and Sharma and George [48] for examples). The second-order system's solution is provided by Hargreaves and Higham [47] utilizing sine and cosine matrices. Sharma and George [48] investigated the controllability of second-order systems.

Lemma 8.3 *[48] The subsequent claims are interchangeable:*

- *The linear system (6.1) is controllable on $[0, \mathscr{T}]$.*
- *The rows of $\mathscr{A}^{-1}\sin(\mathscr{A}t)\mathscr{B}$ are linearly independent.*
- *The controllability Grammian,*

$$\mathscr{W}_0^{\mathscr{T}} = \int_0^{\mathscr{T}} \mathscr{A}^{-1}\sin(\mathscr{A}(\mathscr{T}-s))\mathscr{B}\mathscr{B}^*(\mathscr{A}^{-1}\sin(\mathscr{A}(\mathscr{T}-s)))^* ds,$$

   is non-singular.
- *$Rank[\mathscr{B} : \mathscr{A}^2\mathscr{B} : (\mathscr{A}^2)^2\mathscr{B} : \ldots : (\mathscr{A}^2)^{n-1}\mathscr{B}] = m$.*

Studying the second-order neutral integrodifferential system's controllability is the primary objective of this section.

$$\frac{d}{dt}[z'(t) + g(t,z(t))] + \mathscr{A}^2 z(t) = \mathscr{B}v(t) + f(t,z(t))$$

$$+ \int_0^t h(t,s,z(s))ds, \quad t \in \mathscr{I}, \, t \neq t_j, \tag{8.24}$$

$$z(0) = z_0, \quad z'(0) = x_0, \tag{8.25}$$

where

- $z = z(t) \in \mathscr{R}^n$ is the system vector; and
- $v \in \mathscr{R}^n$ is the control function;
- The constant non-singular matrix $\mathscr{A} \in \mathscr{R}^{n \times n}$ is called the second-order state matrix;
- $\mathscr{B}$ is the constant matrix of order $n \times m$, $\mathscr{I}$ is the interval $[0, \mathscr{T}]$; and
- $g(\cdot), f(\cdot) : \mathscr{I} \times \mathscr{R}^n \to \mathscr{R}^n$, $h(\cdot) : \mathscr{I} \times \mathscr{I} \times \mathscr{R}^n \to \mathscr{R}^n$ are appropriate non-linear functions.

Make a list concerning the Banach space.

$$\mathscr{C}(\mathscr{I}, \mathscr{R}^n) := \{z : \mathscr{I} \to \mathscr{R}^n : x \in C(\mathscr{I}, \mathbb{R}^n)\} \text{ exists}$$

with the norm

$$\|z\|_{\mathscr{C}} := \sup_{t \in \mathscr{I}} \|z(t)\|.$$

The space

$$\mathscr{C}'(\mathscr{I}, \mathbb{R}^n) := \{x \in \mathscr{C}(\mathscr{I}, \mathbb{R}^n) : x \in C'(I, \mathbb{R}^n)\} \text{exists}$$

with the norm $\|z\|_{\mathscr{C}'} := \max\{\|z\|_{\mathscr{P}C}, \|z'\|_{\mathscr{P}C}\}$ is a Banach space.

If $z(\cdot)$ is a solution of Equations (8.24) and (8.25) and $g(t,z(t))$ is sufficiently smooth, then (see Ref. [48]) we get

$$z(t) = \cos(\mathscr{A}t)z_0 + \mathscr{A}^{-1}\sin(\mathscr{A}t)[z_0 + g(0,z(0))] - \int_0^t \cos(\mathscr{A}(t-s))g(s,z(s))ds$$

$$+ \int_0^t \mathscr{A}^{-1}\sin(\mathscr{A}(t-s))\mathscr{B}v(s)ds + \int_0^t \mathscr{A}^{-1}\sin(\mathscr{A}(t-s))f(s,z(s))ds$$

$$+ \int_0^t \mathscr{A}^{-1}\sin(\mathscr{A}(t-s))\int_0^s h(s,\tau,z(\tau))d\tau ds, \quad t \in \mathscr{I}.$$

Using this definition, $z(\cdot)$ is the solution of the problem (8.24) and (8.25) and then, by the properties of second-order differential equation, we have

$$z'(t) = -\mathscr{A}\sin(\mathscr{A}t)z_0 + \cos(\mathscr{A}t)[x_0 + g(0,z(0))] - g(t,z(t)) + \int_0^t \mathscr{A}\sin(\mathscr{A}(t-s))g(s,z(s))ds$$

$$+ \int_0^t \cos(\mathscr{A}(t-s))[\mathscr{B}v(s) + f(s,z(s))]ds + \int_0^t \cos(\mathscr{A}(t-s))\int_0^s h(s,\tau,z(\tau))d\tau ds, \, t \in \mathscr{I}.$$

Definition 8.7 *[38] The system (8.24) and (8.25) is said to be controllable on the interval $\mathscr{I} = [0, \mathscr{T}]$, if for every $(z_0, x_0) \in \mathscr{R}^n \times \mathscr{R}^n$ and $(z_\mathscr{T}, x_\mathscr{T}) \in \mathscr{R}^n \times \mathscr{R}^n$, there exists a control $v \in \mathbb{L}^2(\mathscr{I}, \mathbb{R}^n)$ such that the solution $z(\cdot)$ of Equations (8.24) and (8.25) satisfies $z(\mathscr{T}) = z_\mathscr{T}, z'(\mathscr{T}) = x_\mathscr{T}.$*

Let

$$\mathbb{B}_r := \{z \in \mathbb{R}^n : \|z\| \leq r\}, \quad \text{for some } r \geq 1.$$

The subsequent controllability matrices are required for the validation of the controllability result:

$$\mathscr{W}_1 = \int_0^T \mathscr{A}^{-1} \sin(\mathscr{A}(\mathscr{T}-s))\mathscr{B}\mathscr{B}^*(\mathscr{A}^{-1}\sin(\mathscr{A}(\mathscr{T}-s)))^* ds$$

and

$$\mathscr{W}_2 = \int_0^T \cos(\mathscr{A}(\mathscr{T}-s))\mathscr{B}\mathscr{B}^*(\cos(\mathscr{A}(\mathscr{T}-s)))^* ds.$$

Lets employ the following premises in order to examine the controllability problem:

(H1) Assume that $\mathscr{W}_1^{-1}$, $\mathscr{W}_2^{-1}$ exist and take $K_1 = \|\mathscr{W}_1^{-1}\|$ and $K_2 = \|\mathscr{W}_2^{-1}\|$.

(H2) $\mathscr{A}^{-1}\sin(\mathscr{A}(\mathscr{T}-s))\mathscr{B}\mathscr{B}^*(\cos(\mathscr{A}(\mathscr{T}-s)))^* = \cos(\mathscr{A}(\mathscr{T}-s))\mathscr{B}\mathscr{B}^*(\mathscr{A}^{-1}\sin(\mathscr{A}(\mathscr{T}-s)))^* = 0.$

(H3) The nonlinear function $g : \mathscr{I} \times \mathscr{R}^n \to \mathscr{R}^n$ is continuous, and there exist positive constants $\mathscr{L}_g > 0$, $\widehat{\mathscr{L}_g} > 0$ such that

$$\|g(t,y(t)) - g(s,z(t))\| \leq \mathscr{L}_g[|t-s| + \|y-z\|], \quad \text{for } t,s \in \mathscr{I}, \ y, z \in \mathscr{R}^n,$$

and $\widehat{\mathscr{L}_g} = \max_{t \in \mathscr{I}} \|g(t,0)\|$.

(H4) The function $f : \mathscr{I} \times \mathscr{R}^n \to \mathbb{R}^n$ satisfies the following conditions:

(i) For each $t \in \mathscr{I}$, the function $f(t,.) : \mathscr{R}^n \to \mathscr{R}^n$ is continuous, and for each $z \in \mathscr{R}^n$, the function $f(.,z) : \mathscr{I} \to \mathscr{R}^n$ is strongly measurable.

(ii) There exist positive constants $\mathscr{L}_f > 0$, $\widehat{\mathscr{L}_f} > 0$ such that

$$\|f(t,z_1) - f(t,z_2)\| \leq \mathscr{L}_f \|z_1 - z_2\|, \quad \text{for any } t \in \mathscr{I}, \ z_i \in \mathscr{R}^n, i = 1,2,$$

and $\widehat{\mathscr{L}_f} = \max_{t \in \mathscr{I}} \|f(t,0)\|$.

(H5) The function $h(t,s,.) : \mathscr{R}^n \to \mathscr{R}^n$ is continuous, and for each $z \in \mathscr{R}^n$, the function $h(.,.,z) : \mathscr{I} \times \mathscr{I} \to \mathscr{R}^n$ is strongly measurable. There exist constants $\mathscr{L}_h > 0$, $\widehat{\mathscr{L}_h} > 0$ such that

$$\|h(t,s,y) - h(t,s,z)\| \leq \mathscr{L}_h \|y-z\|, \quad \text{for any } t,s \in \mathscr{I}, \ y,z \in \mathscr{R}^n,$$

and $\widehat{\mathscr{L}_h} = \max_{t \in \mathscr{I}} \|h(t,s,0)\|$.

(H6) There exist constants $r > 0$, $\widehat{r} > 0$ such that

$$M_C\|z_0\| + M_S[\|x_0\| + \widehat{\mathscr{L}_g}] + \mathscr{T}M_C[\mathscr{L}_g r + \widehat{\mathscr{L}_g}] + \mathscr{T}M_S\mathscr{L}_B\mathscr{L}_0$$
$$+ \mathscr{T}M_S[\mathscr{L}_f r + \widehat{\mathscr{L}_f}] + \mathscr{T}M_S[\mathscr{L}_h r + \widehat{\mathscr{L}_h}] \leq r, \quad and$$

$$\tilde{M}_S\|x_0\| + M_C[\|z_0\| + \widehat{\mathscr{L}_g}] + (1 + \mathscr{T}\tilde{M}_S)[\mathscr{L}_g r + \widehat{\mathscr{L}_g}] + \mathscr{T}M_C M_B \mathscr{L}_0$$
$$+ \mathscr{T}M_C[\mathscr{L}_f r + \widehat{\mathscr{L}_f}] + \mathscr{T}M_C[\mathscr{L}_h r + \widehat{\mathscr{L}_h}] \leq \widehat{r}.$$

To make things more readily let's say

$$M_C = \{\|\cos(\mathscr{A}t)\| : t \in [0,\mathscr{T}]\}, \ M_S = \{\|A^{-1}\sin(\mathscr{A}t)\| : t \in [0,\mathscr{T}]\},$$
$$\tilde{M}_S = \{\|\mathscr{A}\sin(\mathscr{A}t)\| : t \in [0,\mathscr{T}]\}, \ M_B = \|B\|, \ M_W = \|\mathscr{W}_1\| \text{ and } \tilde{M}_W = \|\mathscr{W}_2\|.$$

Also let

$$\gamma_1 = (1 + M_W K_1)\left[\mathscr{T}M_C\mathscr{L}_g + \mathscr{T}M_S\mathscr{L}_f + \mathscr{T}^2 M_S\mathscr{L}_h\right]$$

and

$$\gamma_2 = [1 + \tilde{M}_W K_2]\left(\mathscr{L}_g + \mathscr{T}\tilde{M}_S\mathscr{L}_g + \mathscr{T}M_C\mathscr{L}_f + \mathscr{T}^2 M_C\mathscr{L}_h\right).$$

Theorem 8.2 *If the assumptions (H1)–(H6) are satisfied and, if $0 \le \gamma_1, \gamma_2 < 1$, then the system (8.24) and (8.25) is controllable on $\mathscr{I}$.*

Proof. We construct the control function for an arbitrary function $z(\cdot)$ using the matrices $\mathscr{W}_1$ and $\mathscr{W}_2$

$$
\begin{aligned}
v(t) \;=\;& \mathscr{B}^*(\mathscr{A}^{-1}\sin(\mathscr{A}(\mathscr{T}-t)))^*\mathscr{W}_1^{-1}\Big[z_T - \cos(\mathscr{A}t)z_0 - \mathscr{A}^{-1}\sin(\mathscr{A}t)[x_0 + g(0,z(0))] \\
&+ \int_0^{\mathscr{T}}\cos(\mathscr{A}(\mathscr{T}-s))g(s,z(s))ds - \int_0^T A^{-1}\sin(A(T-s))f(s,z(s))ds \\
&- \int_0^b \mathscr{A}^{-1}\sin(\mathscr{A}(\mathscr{T}-s))\int_0^s h(s,\tau,z(\tau))d\tau ds\Big] + \mathscr{B}^*(\cos(\mathscr{A}(\mathscr{T}-t)))^*\mathscr{W}_2^{-1}\Big[x_T + \mathscr{A}\sin(\mathscr{A}t)z_0 \\
&- \cos(\mathscr{A}t)[z_0 + g(0,z(0))] + g(\mathscr{T},z(\mathscr{T})) - \int_0^{\mathscr{T}} A\sin(\mathscr{A}(\mathscr{T}-s))g(s,z(s))ds \\
&- \int_0^{\mathscr{T}}\cos(\mathscr{A}(\mathscr{T}-s))f(s,z(s))ds - \int_0^{\mathscr{T}}\cos(\mathscr{A}(\mathscr{T}-s))\int_0^s h(s,\tau,z(\tau))d\tau ds\Big].
\end{aligned}
$$

One have to demonstrate that when utilizing this control, the operator

$$
\Omega : \mathscr{C}(\mathscr{I},\mathbb{B}_r) \to \mathscr{C}(\mathscr{I},\mathbb{B}_r),
$$

defined by

$$
\begin{aligned}
(\Omega z)(t) \;=\;& \cos(\mathscr{A}t)z_0 + \mathscr{A}^{-1}\sin(\mathscr{A}t)[z_0 + g(0,z(0))] - \int_0^t \cos(\mathscr{A}(t-s))g(s,z(s))ds \\
&+ \int_0^t A^{-1}\sin(A(t-s))\mathscr{B}\Big\{\mathscr{B}^*(\mathscr{A}^{-1}\sin(\mathscr{A}(T-s)))^*\mathscr{W}_1^{-1}\Big[z_T - \cos(\mathscr{A}T)z_0 \\
&- \sin(\mathscr{A}T)[x_0 + g(0,z(0))] + \int_0^T \cos(\mathscr{A}(T-s))g(s,x(s))ds \\
&- \int_0^T \mathscr{A}^{-1}\sin(\mathscr{A}(T-s))[f(s,z(s)) + \int_0^s h(s,\tau,x(\tau))d\tau]ds\Big](s) \\
&+ \mathscr{B}^*(\cos(\mathscr{A}(T-s)))^*\mathscr{W}_2^{-1}\Big[x_T + \mathscr{A}\sin(\mathscr{A}T)z_0 - \cos(AT)[z_0 + g(0,z(0))] \\
&+ g(T,z(T)) - \int_0^T \mathscr{A}\sin(\mathscr{A}(T-s))g(s,z(s))ds \\
&- \int_0^T \cos(\mathscr{A}(T-s))[f(s,z(s)) + \int_0^s h(s,\tau,z(\tau))d\tau]ds\Big](s)\Big\}ds \\
&+ \int_0^t \mathscr{A}^{-1}\sin(\mathscr{A}(t-s))f(s,z(s))ds \\
&+ \int_0^t \mathscr{A}^{-1}\sin(\mathscr{A}(t-s))\int_0^s h(s,\tau,z(\tau))d\tau ds \tag{8.26}
\end{aligned}
$$

has a fixed point $z(\cdot)$ of Equations (8.24) and (8.25). The operator Ω becomes continuous because all of the functions it employs are continuous.

Let's take it for convenience's sake

$$\mathscr{Q}(\eta,z)$$
$$= \mathscr{B}^*(\mathscr{A}^{-1}\sin(\mathscr{A}(T-t)))^*\mathscr{W}_1^{-1}\Big[z_T - \cos(\mathscr{A}T)z_0 - \mathscr{A}^{-1}\sin(\mathscr{A}T)[z_0 + g(0,z(0))]$$
$$+ \int_0^T \cos(\mathscr{A}(T-s))g(s,z(s))ds - \int_0^T \mathscr{A}^{-1}\sin(\mathscr{A}(T-s))f(s,z(s))ds$$
$$- \int_0^b A^{-1}\sin(\mathscr{A}(T-s))\int_0^s h(s,\tau,z(\tau))d\tau ds\Big](\eta) + \mathscr{B}^*(\cos(\mathscr{A}(T-t)))^*\mathscr{W}_2^{-1}\Big[x_T + \mathscr{A}\sin(\mathscr{A}T)z_0$$
$$- \cos(AT)[z_0 + g(0,z(0))] + g(T,z(T)) - \int_0^T \mathscr{A}\sin(\mathscr{A}(T-s))g(s,z(s))ds$$
$$- \int_0^T \cos(\mathscr{A}(T-s))f(s,z(s))ds - \int_0^T \cos(\mathscr{A}(T-s))\int_0^s h(s,\tau,z(\tau))d\tau ds\Big](\eta).$$

From our assumptions, we have

$$\|\mathscr{Q}(\eta,z)\| \leq M_B M_S K_1\Big[\|z_T\| + M_C\|x_0\| + M_S[\|y_0\| + \widehat{\mathscr{K}_g}] + \mathscr{T}M_C[\mathscr{K}_g r + \widehat{\mathscr{K}_g}]$$
$$+ \mathscr{T}M_S[\mathscr{K}_f r + \widehat{\mathscr{K}_f}] + \mathscr{T}^2 M_S[\mathscr{K}_h r + \widehat{\mathscr{K}_h}]\Big] + M_B M_C K_2\Big[\|y_T\| + \widetilde{M}_S\|z_0\|$$
$$+ M_C[\|z_0\| + \widehat{\mathscr{K}_g}] + (1 + \mathscr{T}\widetilde{M}_S)[K_g r + \widehat{\mathscr{K}_g}] + \mathscr{T}M_C[\mathscr{K}_f r + \widehat{\mathscr{K}_f}] + \mathscr{T}^2 M_C[\mathscr{K}_h r + \widehat{\mathscr{K}_h}]\Big]$$
$$= \mathscr{L}_0$$

and

$$\|\mathscr{Q}(\eta,y) - \mathscr{Q}(\eta,z)\| \leq \Big\{M_B M_S K_1\Big(\mathscr{T}M_C\mathscr{K}_g + \mathscr{T}M_S\mathscr{K}_f + \mathscr{T}^2 M_S\mathscr{K}_h\Big)$$
$$+ M_B M_C K_2\Big(K_g + \mathscr{T}\widetilde{M}_S\mathscr{K}_g + \mathscr{T}M_C\mathscr{K}_f + \mathscr{T}^2 M_C\mathscr{K}_h\Big)\Big\}\|y - z\|.$$

First, we show that Ω maps $\mathscr{PC}(\mathbb{I},\mathbb{B}_r)$ into itself. Now

$$\|(\Omega z)(t)\| \leq M_C\|z_0\| + M_S[\|x_0\| + \widehat{\mathscr{L}_g}] + TM_C[L_g r + \widehat{\mathscr{L}_g}] + \mathscr{T}M_B M_S \mathscr{L}_0$$
$$+ \mathscr{T}M_S[\mathscr{L}_f r + \widehat{\mathscr{L}_f}] + \mathscr{T}M_S[\mathscr{L}_h r + \widehat{\mathscr{L}_h}],$$

and

$$\|(\Omega z)'(t)\| \leq \widetilde{M}_S\|z_0\| + M_C[\|z_0\| + \widehat{\mathscr{L}_g}] + (1 + \mathscr{T}\widetilde{M}_S)[\mathscr{L}_g r + \widehat{\mathscr{L}_g}] + \mathscr{T}M_C M_B \mathscr{L}_0$$
$$+ \mathscr{T}M_C[\mathscr{L}_f r + \widehat{\mathscr{L}_f}] + \mathscr{T}M_C[\mathscr{L}_h r + \widehat{\mathscr{L}_h}].$$

From the assumption $(H7)$, $\|(\Omega z)(t)\| \leq r$ and $\|(\Omega z)'(t)\| \leq \hat{r}$. Therefore, Ω maps $\mathscr{PC}(\mathscr{I},\mathscr{B}_r)$ into itself. Moreover, if $y,z \in \mathscr{PC}(\mathscr{I},\mathscr{B}_r)$, then

$$\|(\Omega y)(t) - (\Omega z)(t)\| \leq \Big\{(1 + M_W K_1)\Big[\mathscr{T}M_C\mathscr{L}_g + \mathscr{T}M_S\mathscr{L}_f + \mathscr{T}^2 M_S\mathscr{L}_h\Big]\Big\}\|y - z\|$$
$$= \gamma_1\|y - z\|_{\mathscr{PC}}. \tag{8.27}$$

Also

$$\|(\Omega y)'(t) - (\Omega z)'(t)\| \leq \Big\{[1 + \widetilde{M}_W K_2]\Big(\mathscr{L}_g + \mathscr{T}\widetilde{M}_S\mathscr{L}_g + \mathscr{T}M_C\mathscr{L}_f + \mathscr{T}^2 M_C\mathscr{L}_h\Big)\Big\}\|y - z\|$$
$$= \gamma_2\|y - z\|_{\mathscr{PC}}. \tag{8.28}$$

Since $\gamma_1 < 1$ and $\gamma_2 < 1$, Equations (8.27) and (8.28) demonstrate that the operator Ω remains a contraction on $\mathscr{C}(I,\mathbb{B}_r)$, there exists an exclusive fixed point according to Banach's fixed point theorem. Therefore, $z \in \mathscr{C}(\mathscr{I},\mathbb{B}_r)$ such that $z(\mathscr{T}) = (\Omega z)(\mathscr{T}) = z_T$, $z'(\mathscr{T}) = (\Omega z)'(\mathscr{T}) = x_T$ implying that the system (8.24) and (8.25) is controllable on $\mathscr{I}$. This proves Theorem 8.2. $\qquad\square$

8.7 CONCLUSION

The contribution gives controllability findings over abstract second-order neutral integrodifferential structures in Banach spaces employing strongly continuous cosine families of operators in addition to the Banach fixed point idea. There is additional discussion of the controllability of second-order neutral integro-differential systems within finite dimensional regions. The outcome shows that it can be done to generate adequate conditions in control problems using the Banach fixed point theorem. Obtaining controllability outcomes for specified systems with fractional derivatives could be one of the future aims.

Conflict of interest

The authors declare no conflict of interest.

REFERENCES

1. Adimy, M., & Ezzinbi, K. A. (1998). Class of linear partial neutral functional-differential with nondense domain. *Journal Differential Equations*, 147, 285–332.
2. Balachandran, K., & Anandhi, E. R. (2003). Neutral functional integrodifferential control systems in Banach spaces. *Kybernetika*, 39, 359–367.
3. Balachandran, K., & Anandhi, E. R. (2004). Controllability of neutral functional integrodifferential infinite delay systems in Banach spaces. *Taiwanese Journal of Mathematics*, 8, 689–702.
4. Fu, X. (2003). Controllability of neutral functional differential systems in abstract space. *Applied Mathematics and Computation*, 141, 281–296.
5. Henriquez, H. R., Hernandez, E., & Dos Santos, J. P. C. (2009). Existence results for abstract partial neutral integrodifferential equation with unbounded delay. *Electronic Journal of Qualitative Theory and Differential Equations*, 29, 1–23.
6. Fu, X. (2004). Controllability of abstract neutral functional differential systems with unbounded delay. *Applied Mathematics and Computation*, 141, 299–314.
7. Hernandez, E., & Henriquez, H. R. (2006). Impulsive partial neutral differential equations. *Applied Mathematics Letters*, 19, 215–222.
8. Hernandez, E., Rabello, M., & Henriquez, H. R. (2007). Existence of solutions for impulsive partial neutral functional differential equations. *Journal of Mathematical Analysis and Applications*, 331, 1135–1158.
9. Wang, L., & Wang, Z. (2002). Controllability of abstract neutral functional differential systems with infinite delay. *Dynamics of Continuous, Discrete and Impulsive Systems Series B: Applications and Algorithms*, 9, 59–70.
10. Hale, J. K., & Verduyn Lunel, S. M. (1993). *Introduction to Functional-Differential Equations*. Springer-Verlag, New York.
11. Fitzgibbon, W. (1980). Semilinear integrodifferential equations in Banach spaces. *Nonlinear Analysis: Theory, Methods & Applications*, 4, 745–760.
12. Curtain, R. F., & Zwart, H. (1995). *An Introduction to Infinite Dimensional Linear Systems Theory*. Springer-Verlag, Berlin.
13. Kisynski, J. (1972). On cosine operator functions and one parameter group of operators. *Studia Mathematica*, 49, 93–105.
14. Balachandran, K. (1989). Controllability of nonlinear Volterra integrodifferential systems. *Kybernetika*, 25, 505–508.
15. Balachandran, K., & Dauer, J. P. (1987). Controllability of nonlinear systems via fixed point theorems. *Journal of Optimization Theory and Applications*, 53, 345–352.
16. Klamka, J. (1993). *Controllability of Dynamical Systems*. Kluwer Academic, Dordrecht.
17. Pazy, A. (1983). *Semigroups of Linear Operators and Applications to Partial Differential Equations*. Springer-Verlag, New York.
18. Balachandran, K., & Dauer, J. P. (2002). Controllability of nonlinear systems in Banach spaces: A survey. *Journal of Optimization Theory and Applications*, 115, 7–28.
19. Balachandran, K., & Sakthivel, R. (2000). Controllability of semilinear functional integrodifferential systems in Banach spaces. *Kybernetika*, 36, 465–476.

20. Balachandran, K., Park, J. Y., & Anthoni, M. (1999). Controllability of second order semilinear Volterra integrodifferential systems in Banach spaces. *Bulletin of the Korean Mathematical Society*, 36, 1–13.
21. Bochenek, J. (1991). An abstract nonlinear second-order differential equation. *Annales Polonici Mathematici*, 54, 155–166.
22. Bochenek, J. (1992). Second order semilinear integrodifferential equation. *Annales Polonici Mathematici*, 52, 231–241.
23. Fattorini, H. O. (1981). Some remarks on second order abstract Cauchy problems. *Funkcialaj Ekvacioj*, 24, 331–344.
24. Fattorini, H. O. (1985). *Second Order Linear Differential Equations in Banach spaces*. Elsevier, North Holland, Amsterdam.
25. Henriquez, H. R. (1985). On non-exact controllable systems. *International Journal of Control*, 42, 71–83.
26. Park, J. Y., & Han, H. K. (1997). Controllability for some second order differential equations. *Bulletin of the Korean Mathematical Society*, 34, 411–419.
27. Travis, C. C., & Webb, G. F. (1977). Compactness, regularity and uniform continuity properties of strongly continuous cosine families. *Houston Journal of Mathematics*, 3, 555–567.
28. Travis, C. C., & Webb, G. F. (1978). Cosine families and abstract nonlinear second order differential equations. *Acta Mathematica Academiae Scientiarum Hungarica*, 32, 75–96.
29. Travis, C. C., & Webb, G. F. (1979). An abstract second order semi-linear Volterra integro-differential equation. *SIAM Journal of Mathematical Analysis*, 10, 412–424.
30. Randelovic, B. M., Stefanovic, L. V., & Dankovic, B. M. (2000). Numerical solution of impulsive differential equations. *Series Mathematics and Informatics*, 15, 101–111.
31. Heard, M. L. (1984). A quasilinear hyperbolic integro-differential equation related to a nonlinear string. *Transactions of American Mathematical Society*, 285, 805–823.
32. Matos, M., & Pereira, D. (1991). On a hyperbolic equation with strong damping. *Funkcialaj Ekvacioj*, 34, 303–311.
33. Patcheu, S. K. (1997). On the global solution and asymptotic behavior for the generalized damped extensible beam equation. *Journal of Differential Equations*, 135, 299–314.
34. Triggiani, R. (1975). On the lack of exact controllability for mild solutions in Banach spaces. *Journal of Mathematical Analysis and Applications*, 50, 438–446.
35. Hernandez, E., & Henriquez, H. R. (2009). Existence results for second order differential equations with non-local conditions in Banach spaces. *Funkcialaj Ekvacioj*, 52, 113–137.
36. Hernandez, E., Rabello, M., & Henriquez, H. R. (2007). Existence of solutions for impulsive partial neutral functional differential equations. *Journal of Mathematical Analysis and Applications*, 331, 1135–1158.
37. Balachandran, K., & Kim, J. H. (2006). Remarks on the paper controllability of second order differential inclusion in Banach spaces [*Journal of Mathematical Analysis and Applications*, 285, 537–555 (2003)]. *Journal of Mathematical Analysis and Applications*, 324, 746–749.
38. Henriquez, H. R. (1985). On non exact controllable systems. *International Journal of Control*, 42, 71–83.
39. Granas, A., & Dugundji, J. (2003). *Fixed Point Theory*. Springer-Verlag, New York.
40. Martin, R. H. (1987). *Nonlinear Operators and Differential Equations in Banach Spaces*. Robert E. Krieger Publishing Company, Malabar, FL.
41. Byszewski, L. (1991). Theorems about the existence and uniqueness of solutions of a semi-linear evolution non-local Cauchy problem. *Journal of Mathematical Analysis and Applications*, 162, 494–505.
42. Byszewski, L. (1997). On a mild solution of a semilinear functional-differential evolution non-local problems. *Journal of Applied Mathematics and Stochastic Analysis*, 10, 265–271.
43. Byszewski, L., & Akca, H. (1998). Existence of solutions of a semi-linear functional-differential evolution non-local problem. *Nonlinear Analysis*, 34, 65–72.
44. Balachandran, K., & Park, J. Y. (2001). Existence of solution of second order nonlinear differential equations with non-local conditions in Banach spaces. *Indian Journal of Pure and Applied Mathematics*, 32, 1883–1892.
45. Han, H. K., & Park, J. Y. (1999). Boundary controllability of differential equations with non-local conditions. *Journal of Mathematical Analysis and Applications*, 230, 241–250.
46. Hughes, P. C., & Skelton, R.E. (1980). Controllability and observability of linear matrix second order systems. *ASME Journal of Applied Mechanics*, 47, 415–424.

47. Hargreaves, G. I., & Higham, N. J. (2005). Efficient algorithms for the matrix Cosine and Sine. *Numerical Algorithms*, 40, 383–400. DOI: 10.1007/s11075-005-8141-0.
48. Sharma, J. P., & George, R. K. (2007). Controllability of matrix second order systems. *Electronic Journal of Differential Equations*, 80, 1–14.

9 Fuzzy Random Multi-Objective Quadratic Transportation Problem

P. Anukokila and B. Radhakrishnan

9.1 INTRODUCTION

The main trend in science and engineering around the turn of the 20th century was the distillation of complex systems in reality into precise mathematical models. Operations research (OR) was initiated in the middle of this century to apply the real-world decision-making issues. It was developed as a result of World War II military planning. After the war, the methods started to be used more often to address issues in business, industry, and society. Since then, operations research has developed into a discipline that is widely used in a variety of sectors, including finance, petrochemicals, and aviation. It has also moved to a focus on the creation of mathematical models that can be used to analyze and optimize complex systems. Operations research is the methodical use of mathematical and statistical approaches to address issues in industry, government, and education.

In Applied Mathematics, optimization is the central concept. Several disciplines are extremely interested in it. Mathematical optimization may be used to formulate many modeling, design, control, and decision-making issues. The minimization (or maximization) of the objectives, in light of the restrictions imposed by the issue at hand, constitutes the traditional framework for optimization. However, many design issues involve a number of objectives that must be traded off, which can result in the under- or over-achievement of numerous goals. Principle of optimization theory has been proposed by Bector, Chandra, and Dutta [5]. Optimizing techniques such as linear programming have proved useful in solving limited kinds of operating problems but have not been widely used to tackle higher level management problems. The availability of time-sharing services has permitted the use of non-optimizing techniques (e.g., simulation) to solve a wider range of problems than would be possible with analytic models.

Numerous system performance characteristics, criteria, and decision-making factors cannot always be properly quantified, nor is it always necessary. When a variable's values cannot be accurately established, it is referred to as uncertain or fuzzy. Uncertainty in values can be quantified using probability distributions. Alternatively, if they are better described by qualitative descriptors like dry or wet, hot or cold, clean or filthy, and high or low, fuzzy membership functions can be employed to quantify them. For these ambiguous or qualitative variables, quantitative optimization models can incorporate both fuzzy membership functions and probability distributions. For the purpose of solving real-world decision-making issues, fuzzy set theory is a helpful tool. It has been utilized in a number of OR strategies, including queuing theory, nonlinear programming, and linear programming. This study will focus on linear programming, the most crucial approach out of the bunch.

Zadeh, on the other hand, coined the term "fuzzy" in 1962 and formally published the classic article "Fuzzy Set Theory" in 1965 [52]. The oversimplified model was intended to be enhanced using fuzzy set theory in order to provide a more robust and adaptable model that could be applied to complicated human-centric systems in the real world. It is now widely accepted the fact that fuzzy structures and fuzzy theories of control have some advantages over numerous other methods

DOI: 10.1201/9781003407386-9

whenever an intricate physical system fails to offer a set of differential or difference equations that constitute a reliable or fairly precise mathematical framework, in addition to when the system description necessitates a certain level of human comprehension in terms of language. Only in the twentieth century were the concepts of sets and functions created by mathematicians to express problems. This issue representation approach is stricter. The solutions based on this theory are frequently worthless. This issue was resolved with the aid of the fuzzy idea. Almost all mathematical, engineering, medical, and other concepts have been redefined using fuzzy sets. Therefore, it is crucial to spread awareness of these principles for our future generation.

9.2 BACKGROUND AND RELATED WORK

9.2.1 FUZZY SET

A fuzzy set is one that has "un-sharp" and ambiguous bounds. It transforms the concept of membership from a binary categorization in classical set theory to one that admits partial membership. The constraints of classical set theory are overcome by fuzzy set theory, which allows membership in a set to be a question of degree. A value between 0 and 1 represents the degree of membership in a fuzzy set; 0 indicates fully not in the set, and 1 implies completely in the set.

In one of his earliest publications on fuzzy set theory, Zadeh writes, "Ultimately, the principle of a fuzzy set serves as an ideal point concerning departure for the creation of a theoretical structure which parallels in many ways the model used in the case of conventional sets, but is more general than the former and, potentially, may prove to have an even wider range of feasibility, notably in the domains of pattern categorization and information processing." Essentially, such a framework "provides a logical approach of dealing with situations when the cause of imprecision is the absence of finely specified class membership requirements rather than the presence of random variables." The book written by Zimmermann [55] performed an important contribution in the development of the field known as fuzzy sets decision-making and expert systems from the perspective of application in science and engineering.

Applications of this theory include artificial intelligence, computer science, healthcare, control engineering, decision theory, expert systems, logic, management science, operations research, pattern recognition, and robotics. Mathematical advancements have progressed to a very high level and continue to do so now. This topic focuses on optimization theory, with a particular emphasis on fuzzy linear programming. Despite the existence of many alternative techniques such as queuing theory, heuristic approaches, nonlinear programming, and so on, a review of OR literature reveals that linear programming is the most essential technique for real-world applications. The methods and applications of fuzzy mathematical programming have been established by Zimmermann [54]. Moreover, he studied [56] the optimization of fuzzy systems.

9.2.2 LINEAR PROGRAMMING

The phrase "linear optimization" can also refer to linear programming. Simply put, linear programming involves utilizing a mathematical model that is linear to optimize an output based on a set of constraints. The following symbolically expresses the generic LPP:

$$\begin{cases} \text{maximize} \quad \mathscr{Z} = p^T y \\ \text{subject to} \\ \quad \mathscr{A} y \leq d \\ \quad y \geq 0 \end{cases}$$

where p is the vector of profit coefficients of the objective function, d is the vector of total resources available, y is the vector of decision variables, and $\mathscr{A}$ is the matrix of technical coefficients.

In the case of linear programming, an anticipated result, such as profit maximization or cost minimization, is attained by wisely allocating scarce resources to various tasks. All of the interactions between the processes in models of linear programming are linear, satisfying both the proportionality and additive behavior criteria. Contrary to fuzzy linear programming issues, sadly, little study has been done on this significant class of problems up to this point.

One of the most reliable optimization methods used in the planning of reservoir operations is linear programming. Under particular conditions and assumptions, it can linearize nonlinear equations completely, linearize them in pieces, or solve nonlinear problems iteratively. The planning objectives often depend on how much is stored and released within the planning horizon. Most reservoir optimization problems have one of two planning objectives: either increase net management revenue or reduce operating costs while meeting all requirements. The following are the main benefits of linear programming models:

- Ability to deal with intricate problems.
- Convergence to the universal optimum is guaranteed.
- The user does not require any information to make an initial guess.
- Proficiency in carrying out sensitivity analyzes.
- The utmost simplicity of problem formulation.
- Software programmes that are conveniently accessible and effective, such as Lindo and Lingo.

9.2.3 NONLINEAR PROGRAMMING

The process of addressing an optimization issue specified by a collection of constraints, also known as equalities and inequalities, over a number of unknown real variables, together with an objective function that should be maximized or reduced, is known as nonlinear programming (NLP). Over the range of choice variables being taken into consideration, using assumptions or approximations from linear programming may also result in suitable issue representations. The objective of a universal optimization challenge is to choose n choice variables $y_1, y_2, \ldots, y_n$ in the most feasible region while optimizing the objective function representing the decision factors

$$g(y_1, y_2, \ldots, y_n).$$

If the goal function is nonlinear and/or nonlinear constraints are used to define the viable region, the issue is referred to as a nonlinear programming problem (NLPP). As a result, the generic nonlinear programming is expressed as follows in maximization form:

$$\begin{cases} \text{maximize } g(y_1, y_2, \ldots, y_n) \\ \text{subject to} \\ h_1(y_1, y_2, \ldots, y_n) \leq d_1 \\ \quad \vdots \\ h_m(y_1, y_2, \ldots, y_n) \leq d_m \end{cases}$$

where each constraint function spanning h_1 through h_m is provided. The linear program that has already been addressed is an exception. In this instance, the apparent relationship is

$$\begin{cases} g(y_1, y_2, \ldots, y_n) = \sum_{j=1}^{n} p_j y_j \\ \text{and} \\ h_i(y_1, y_2, \ldots, y_n) = \sum_{j=1}^{n} e_{ij} y_j, \ (i = 1, 2, \ldots, m). \end{cases}$$

9.2.4 FUZZY LINEAR AND NONLINEAR PROGRAMMING

According to the crisp scenario, the LPP's targeting is to maximize or minimize an objective function that is linear subject to linear constraints. The decision-maker may, however, only be able to express the objective and/or constraint functions in a "fuzzy sense," as opposed to precisely, in many real-world situations. It is preferable to utilize a fuzzy linear programming form of modeling in these circumstances. As a way to employ mathematical concepts and associated strategies to military programming and organizing problems, Dantzig suggested that "the interrelations between the activities of a large organization could be considered as a linear programming kind model and the optimizing programme achieved by minimizing a linear objective function."

The design is possible to express a universal fuzzy linear programming problem (FLPP) formally in the following manner:

$$\begin{cases} \text{maximize} & \widehat{\mathscr{Z}_{\mathscr{R}}} = \tilde{p}y \\ \text{subject to} & \\ & \mathscr{A}y = d \\ & y \geq 0 \end{cases}$$

where $d \in \mathscr{R}^m$, $y \in \mathscr{R}^n$, $\tilde{p} \in (\mathscr{F}(\mathscr{R}))^n$, $\mathscr{A} \in \mathscr{R}^{m \times n}$, $\mathscr{R}$ is linear ranking function. Several authors [7, 8, 11, 13, 15, 21, 25, 32, 33, 35, 37] have studied the fuzzy linear programming problem.

There are lot of intriguing real-world challenges where the objective function or some of the restrictions may not be linear. Fuzzy nonlinear programming (FNLP) issues are optimization problems with nonlinearities. Abo-Sinna [2] suggested nonlinear multi-objective decision-making under fuzziness. Xie and Jia [49] established the nonlinear fixed charge transportation problem by minimum cost flow-based genetic algorithm. For an appropriate representation of an application as a mathematical program, nonlinearities in the form of nonlinear objective functions or nonlinear constraints are essential at other times. Following are some ways to approach a nonlinear programming issue using fuzzy coefficients:

$$\begin{cases} \text{maximize} & g(\tilde{c}_1, \tilde{c}_2, \ldots, \tilde{c}_k, y) \\ \text{subject to} & \\ & h_i(\tilde{e}_1, \tilde{e}_2, \ldots, \tilde{e}_k, y) \leq \tilde{d}_i \\ & y_i \in [l_k, u_k], \ k = 1, 2, \ldots, n, \ l_k \geq 0 \end{cases}$$

where $y = (y_1, y_2, \ldots, y_n) \in \mathscr{R}^n$ is a n - dimensional real-valued parameter vector, $[l_k, u_k] \subset \mathscr{R}(k = 1, 2, \ldots, n)$, $g(y)$, $h_i(y)$ are continuous arbitrary functions.

9.2.5 TRANSPORTATION PROBLEM

Transporting goods from sources to destinations is the focus of the transportation issue, a specific kind of linear programming problem. Hitchcock is credited with creating the fundamental transportation problem in Ref. [18]. The reason it is known as a transportation problem is because it was initially developed for figuring out the best shipping strategy. Transporting a specific product from each of m origins $i = 1, \ldots, m$ to any of n destinations $k = 1, 2, \ldots, n$ is a typical and well-known transportation problem. The initial locations are manufacturing facilities with the corresponding capacity $(c_1, c_2, \ldots, c_m)$, whereas the ending points are warehouses having the necessary levels of demand $(e_1, e_2, \ldots, e_n)$. Without sacrificing generality, one can suppose that the cost p_{ik} to transport one unit of the product under consideration from the ith source to the kth destination will be $p_{ik} \geq 0$, for all i, k.

Consequently, the standard transportation problem has the following mathematical form:

$$
\begin{cases}
\text{minimize} \;\; \mathscr{Z} = \sum_{i=1}^{m}\sum_{k=1}^{n} p_{ik} y_{ik} \\[2mm]
\text{subject to} \\[1mm]
\sum_{k=1}^{n} y_{ik} = c_i, \quad i = 1,2,\ldots,m \\[2mm]
\sum_{i=1}^{m} y_{ik} = e_k, \quad k = 1,2,\ldots,n \\[2mm]
y_{ik} \geq 0, \quad \forall \; i \text{ and } k \\[2mm]
\sum_{i=1}^{m} c_i = \sum_{k=1}^{n} e_k.
\end{cases}
$$

The basic goal of the transportation problem is to choose a shipment schedule that minimizes the overall shipping expense while meeting supply and demand constraints. Several authors investigated the study of transportation problem [1, 3, 10, 12, 14, 17, 23, 34, 40, 44, 45]. There has been a good deal of study in the literature to find the best answer for balanced transportation challenges.

But in actual circumstances, the decision-maker must deal with an unbalanced transportation issue where the whole supply is less than the total demand. The imbalanced fuzzy transportation issue that Kumar and Kaur formulated in Ref. [26]. The transportation issue may be solved using the simplex technique. However, an alternative approach may be developed to resolve these issues due to the unique nature of the transportation problem. It was suggested by [9, 24] to take a fuzzy approach to the transportation issue.

The transportation issue may be divided into linear and nonlinear categories depending on the make-up of the cost function. An illustration of a network optimization issue is transportation. With the aim of reducing overall distribution costs, it deals with the effective distribution (movement) of product (goods and services) from various supply locations (sources) with a constrained supply to various demand locations (destinations) with a defined demand. The nonlinear transportation problem is presented as follows in meticulous:

- An array of n commodity suppliers with acknowledged capacity for supply and an array of m destinations with known demand.
- The nonlinear, differentiable transportation cost function spanning a unit of something from each source to each destination.

To satisfy demand at each destination while reducing overall transportation costs, we must determine the quantity of goods that must be provided from each source (and maybe sold).

Physical product distribution, often known as transportation issues, has been one of the most significant and fruitful applications of quantitative analysis in solving business difficulties. Basically, the goal is to keep the cost of moving products from one place to another as low as possible while yet meeting each arriving area's demands and allowing each shipping facility to operate to its fullest potential. Network flow issues make up a sizable portion of the practical applications of linear programming.

9.2.6 MULTI-OBJECTIVE TRANSPORTATION PROBLEM

In actual circumstances, there are other transit issues at play. This chapter considers transportation issues that are characterized by several objective functions. The phrase "multi-objective transportation problem" (MOTP) designates a particular class of LPP when all of the objectives are in conflict with one another and the constraints are of the equality type. Comparable to a normal transportation issue with an MOTP, merchandise needs to be delivered from m sources to n destinations, with each having capabilities of $c_1, c_2, \ldots, c_m$ and $e_1, e_2, \ldots, e_n$. Transporting a unit of goods from i^{th} source to k^{th} destination entails a penalty of $p_i k$ as well. Cost, delivery, or other factors could constitute the penalty. A variable y_{ik} represents the unknown quantity to be shipped from i^{th} source to k^{th} destination. A mathematical model of MOTP with w objectives, m sources, and n destinations can be written as:

$$
\left\{
\begin{array}{l}
\text{minimize} \quad \mathcal{Z}_w = \sum_{i=1}^{m} \sum_{k=1}^{n} p_{ik}^{w} y_{ik}, \quad w = 1, 2, \ldots, q \\[2mm]
\text{subject to} \\[1mm]
\quad \sum_{k=1}^{n} y_{ik} \leq c_i, \quad i = 1, 2, \ldots, m \\[2mm]
\quad \sum_{i=1}^{m} y_{ik} \leq e_k, \quad k = 1, 2, \ldots, n \\[2mm]
\quad\quad y_{ik} \geq 0, \quad \text{for all } i \text{ and } k.
\end{array}
\right.
$$

The subscript on $\mathcal{Z}_w$ and superscript on p_{ik}^{w} are related to the wth penalty criterion. It may be assumed that $c_i \geq 0$ and $e_k \geq 0$, for all i, k and the equilibrium condition $\sum_{i=1}^{m} c_i = \sum_{k=1}^{n} e_k$ is satisfied. Several papers [6,41,42,45] appeared on various types of multi-objective transportation problem with fuzzy parameters in fuzzy approach. For figuring out MOTP, there are numerous distinct ways, including geometric programming, interactive fuzzy multi-objective linear programming, fuzzy goal programming, and fuzzy programming methodology. Nevertheless, the notion of fuzzy goal programming strategy has been emphasized in this study.

9.2.7 SOLID TRANSPORTATION PROBLEM

Haley originally introduced the concept of the solid transportation problem (STP) in Ref. [19] and explored the multi-index problem in Ref. [20], which takes into account three different types of constraints: source constraint, destination constraint, and conveyance constraint. As there is only one vehicle, the STP degenerates toward the conventional transportation problem. It is regarded as a unique instance of LPP. The fact that numerous industrial problems take on this particular shape makes it necessary to take this particular type of transportation problem into consideration. It should be emphasized that when a variety of conveyances are available for the shipment of products, the STP becomes necessary. By focusing on a single mode of transportation, the STP can be transformed into a traditional transportation problem. It has been extensively studied by several authors [22, 29, 39, 51].

Take the following aspects of STP into consideration: The decision-maker of a transportation firm will calculate the transportation quantity based on preliminary research in order to break into an unfamiliar market. The decision-maker frequently calculates the shipping capacity as a sufficiently high figure to guarantee the supply out of concern for the company's reputation. A universe, symbolized by the symbol d_k, can be thought of as an adequately huge integer. Occasionally, the decision-maker is not involved in the detailed analysis or actual implementation of the transit plan.

9.2.8 QUADRATIC PROGRAMMING

Quadratic programming constitutes one of the most significant techniques in operations research. An optimization circumstance with a quadratic objective function and a linear restriction is known as a quadratic programming problem. Quadratic programming is sometimes recognized as a separate subject due to its wide range of applications. The cost per unit of a product sent from a given source to a given destination is constant in the linear transportation issue (also known as the "ordinary transportation problem"), regardless of the quantity shipped. The mileage (distance) between every source and every destination is typically assumed to be constant.

The subsequent is an assortment of quadratic programming with fuzzy numbers (QPFN) problems:

$$\begin{cases} \text{minimize } \frac{1}{2}y^T \widetilde{\mathscr{P}}y + y^T \widetilde{\mathscr{D}} \\ \text{subject to} \\ \qquad \mathscr{A}y \leq \tilde{d}. \end{cases}$$

A mathematical modeling strategy called quadratic programming aims to maximize the use of finite resources. Quadratic programming models are typically created in real-world applications to determine a plan of action for the future. Making a forecast of the future circumstances would be employed to calculate the parameter values, which entails a certain amount of risk. It has produced a variety of intriguing uses as well as various beneficial outcomes. Typically, a second-order quadratic system is used to approximate optimization problems with nonlinear objective functions, and then, a normal quadratic programming approach is used to solve them sequentially. Several authors [14, 47, 48] studied goal programming problems.

One of the effective methods for boosting the effectiveness and productivity of commercial enterprises and governmental agencies is quadratic programming. When real situations are considered, vagueness appears in a natural way, and hence, it makes perfect sense to think of fuzzy quadratic programming problems. Igeartaigh [36] proposed fuzzy transportation algorithm, and Lau et al. [27] proposed multi-objective evolutionary algorithm for transportation problem. The context of their discussion of different linear programming models incorporates fuzzy random variables together with fuzzy random coefficients, as Qiao and Wang point out [46]. Fortunately, we have a streamlined simplex technique that is particularly effective at solving such transportation problems.

Quadratic programming can be viewed both as a special case of the nonlinear programming and as a generalization of the linear programming. Pal and Moitra [38] presented a fuzzy goal programming procedure for solving quadratic bi-level programming. Duality in fuzzy quadratic programming was derived by Gupta et al. [16]. Yager developed the scale setting and broader concept for interval-valued fuzzy sets and their application in uncertainty measures in his work reported in Ref. [50].

The fuzzy quadratic programming task is converted into two two-level mathematical programs built around Zadeh's extension principle, [53], to determine the upper and lower bounds of the objective value at possibility level β. Enumerating various β values leads to the numerical derivation of the membership function of the fuzzy objective value. Further, Liu [31] proposed an effective solution method to solving a class of fuzzy quadratic programming problems. Punnen and Zhang [43] provided quadratic bottleneck problems. Numerical solution method for quadratic programming was discussed by the authors of Refs. [28, 30]. Bellman and Zadeh [4] put out the maximization decision notion. By incorporating this idea of fuzzy sets, Zimmermann [57] became the first to do so in mathematical programs.

9.3 PRELIMINARIES

We need some basic definitions and properties of fuzzy random variables related to this chapter. Let $\mathscr{F}_0(\mathscr{R})$ stands for the collection of all compact fuzzy numbers over $\mathscr{R}$.

Definition 9.1 *For* $a_\beta = [a_\beta^{\mathscr{L}}, a_\beta^{\mathscr{U}}]$ *and* $b_\beta = [b_\beta^{\mathscr{L}}, b_\beta^{\mathscr{U}}]$, *we define*

- $a_\beta + b_\beta = [a_\beta^{\mathscr{L}}, a_\beta^{\mathscr{U}}] + [b_\beta^{\mathscr{L}}, b_\beta^{\mathscr{U}}] = [a_\beta^{\mathscr{L}} + b_\beta^{\mathscr{L}}, a_\beta^{\mathscr{U}} + b_\beta^{\mathscr{U}}]$
- $a_\beta - b_\beta = [a_\beta^{\mathscr{L}}, a_\beta^{\mathscr{U}}] - [b_\beta^{\mathscr{L}}, b_\beta^{\mathscr{U}}] = [a_\beta^{\mathscr{L}} - b_\beta^{\mathscr{L}}, a_\beta^{\mathscr{U}} - b_\beta^{\mathscr{U}}]$

- $a_\beta \cdot b_\beta = [a_\beta^{\mathscr{L}}, a_\beta^{\mathscr{U}}] \cdot [b_\beta^{\mathscr{L}}, b_\beta^{\mathscr{U}}] = [a_\beta^{\mathscr{L}} b_\beta^{\mathscr{L}} \wedge a_\beta^{\mathscr{L}} b_\beta^{\mathscr{U}} \wedge a_\beta^{\mathscr{U}} b_\beta^{\mathscr{L}} \wedge a_\beta^{\mathscr{U}} b_\beta^{\mathscr{U}}, a_\beta^{\mathscr{L}} b_\beta^{\mathscr{L}} \vee a_\beta^{\mathscr{L}} b_\beta^{\mathscr{U}} \vee a_\beta^{\mathscr{U}} b_\beta^{\mathscr{L}} \vee a_\beta^{\mathscr{U}} b_\beta^{\mathscr{U}}]$.

- *The order relation " $\leq$ " is defined by*

$$[a_\beta^{\mathscr{L}}, a_\beta^{\mathscr{U}}] \leq [b_\beta^{\mathscr{L}}, b_\beta^{\mathscr{U}}] \ \textit{if and only if} \ a_\beta^{\mathscr{L}} \leq b_\beta^{\mathscr{L}}, \ a_\beta^{\mathscr{U}} \leq b_\beta^{\mathscr{U}}.$$

- *Let $[a_{\beta i}^{\mathscr{L}}, a_{\beta i}^{\mathscr{U}}] \subset \mathscr{R}$, $i \in \mathscr{I}$, $\mathscr{I}$ is the index set, then*

$$\wedge_{i \in \mathscr{I}} [a_{\beta i}^{\mathscr{L}}, a_{\beta i}^{\mathscr{U}}] = [\wedge_{i \in \mathscr{I}} a_{\beta i}^{\mathscr{L}}, \wedge_{i \in \mathscr{I}} a_{\beta i}^{\mathscr{U}}] \ \textit{if} \ \wedge_{i \in \mathscr{I}} a_{\beta i}^{\mathscr{L}} > -\infty.$$
$$\vee_{i \in \mathscr{I}} [a_{\beta i}^{\mathscr{L}}, a_{\beta i}^{\mathscr{U}}] = [\vee_{i \in \mathscr{I}} a_{\beta i}^{\mathscr{L}}, \vee_{i \in \mathscr{I}} a_{\beta i}^{\mathscr{U}}] \ \textit{if} \ \vee_{i \in \mathscr{I}} a_{\beta i}^{\mathscr{L}} > \infty.$$

Definition 9.2 *Let $\{\tilde{a}_i | i \in \mathscr{I}\} \subset \mathscr{F}_0(\mathscr{R})$, $\beta \in (0,1]$, then*

- $\tilde{g} = \wedge_{i \in \mathscr{I}} \tilde{a}_i$ *is defined by a fuzzy number $\tilde{a} \in \mathscr{F}_0(\mathscr{R})$ such that $g_\beta = \wedge_{i \in \mathscr{I}} (a_i)_\beta$.*
- $\tilde{h} = \wedge_{i \in \mathscr{I}} \tilde{a}_i$ *is defined by a fuzzy number $\tilde{a} \in \mathscr{F}_0(\mathscr{R})$ such that $h_\beta = \wedge_{i \in \mathscr{I}} (a_i)_\beta$.*

Definition 9.3 *Let $(\Omega, \mathscr{A}, \mathscr{P})$ be a probability measure space. A mapping $\tilde{a} : \Omega \to \mathscr{F}_0(\mathscr{R})$ is called a fuzzy random variable on $(\Omega, \mathscr{A})$ if, for any $\beta \in (0,1]$*

$$a_\beta(\eta) = \{y | y \in \mathscr{R}, \tilde{a}(\eta)(y) \geq \beta\} = [a_\beta^{\mathscr{L}}(\eta), a_\beta^{\mathscr{U}}(\eta)]$$

is a random interval, that is $a_\beta^{\mathscr{L}}(\eta)$, $a_\beta^{\mathscr{U}}(\eta)$ are two random variables on $(\Omega, \mathscr{A})$. Denote the set of all fuzzy random variables on $(\Omega, \mathscr{A})$ by $\mathscr{FR}(\Omega)$ (where $\eta \in \Omega$). The algebraic operation $*$ on $\mathscr{FR}(\Omega)$ may be defined by

$$(\tilde{u} * \tilde{v})(\eta) = \tilde{u}(\eta) * \tilde{v}(\eta), \textit{ for any } \eta \in \Omega.$$

Also, for any $\tilde{u}, \tilde{v} \in \mathscr{FR}(\Omega)$

$$(\tilde{u} \leq \tilde{v})(\eta) = \tilde{u}(\eta) \leq \tilde{v}(\eta), \textit{ for any } \eta \in \Omega.$$

9.4   PROBLEM FORMULATION

9.4.1   INTERVAL QUADRATIC TRANSPORTATION PROBLEM

This section describes the linear programming model for MOTP in interval numbers including crisp functions. Transporting a unit of the product from source i toward destination j through the conveyance k is accompanied by a fine or cost of $v_{ijk} \geq 0$.

Using the fuzzy random multi-objective quadratic transportation issue described below as an instance

$$\begin{cases} \text{minimize } \widehat{\mathscr{Z}^v}(\tilde{y}) = \sum_{i=1}^{m} \sum_{j=1}^{n} p_{ij} y_{ij} + \frac{1}{2} \sum_{i=1}^{m} \sum_{j=1}^{n} \sum_{k=1}^{l} v_{ijk} y_{ijk} \\ \text{subject to} \\ \sum_{j=1}^{n} \sum_{k=1}^{l} y_{ijk} = \mathscr{A}_i, \ \text{ for } \ i = 1, 2, \ldots, m \\ \sum_{i=1}^{m} \sum_{k=1}^{l} y_{ijk} = \mathscr{B}_j, \ \text{ for } \ j = 1, 2, \ldots, n \\ \sum_{i=1}^{m} \sum_{j=1}^{n} y_{ijk} = \mathscr{E}_k, \ \text{ for } \ k = 1, 2, \ldots, l \\ \quad\quad y_{ijk} \geq 0, \text{ for all } \ i, j, k \end{cases} \quad (9.1)$$

where $p_{ij} \in [p_{ij}^{\mathcal{U}}, p_{ij}^{\mathcal{L}}]$, $\mathcal{A}_i \in [\mathcal{A}_i^{\mathcal{L}}, \mathcal{A}_i^{\mathcal{U}}]$, $\mathcal{B}_j \in [\mathcal{B}_j^{\mathcal{L}}, \mathcal{B}_j^{\mathcal{U}}]$, and $\mathcal{E}_k = [\mathcal{E}_k^{\mathcal{L}}, \mathcal{E}_k^{\mathcal{U}}]$ are the interval inter parts of p_{ij}, $\mathcal{A}_i$, $\mathcal{B}_j$, $\mathcal{E}_k$, respectively. For equality constraint of the "=" form. Locating the upper and lower bound of Equation $(9.1)'s$ objective values is sufficient to determine the range containing the objective values. The values p_{ij}, $\mathcal{A}_i$, $\mathcal{B}_j$, $\mathcal{E}_k$ attain the largest value for $\mathcal{Z}$ can be determined from the following two-level mathematical programming model

$$
\left\{
\begin{aligned}
&\mathcal{Z}_v^{\mathcal{U}} = \text{maximize}_{(p,\mathcal{A},\mathcal{B},\mathcal{E}) \in s} \ \text{minimize}_y \ \mathcal{Z} \\
&\qquad = \sum_{i=1}^{m} \sum_{j=1}^{n} p_{ij} y_{ij} + \frac{1}{2} \sum_{i=1}^{m} \sum_{j=1}^{n} \sum_{k=1}^{l} v_{ijk} y_{ijk} \\
&\text{subject to} \\
&\qquad \sum_{j=1}^{n} \sum_{k=1}^{l} y_{ijk} = \mathcal{A}_i, \quad \text{for} \quad i = 1, 2, \ldots, m \\
&\qquad \sum_{i=}^{m} \sum_{k=1}^{l} y_{ijk} = \mathcal{B}_j, \quad \text{for} \quad j = 1, 2, \ldots, n \\
&\qquad \sum_{i=1}^{m} \sum_{j=1}^{n} y_{ijk} = \mathcal{E}_k, \quad \text{for} \quad k = 1, 2, \ldots, l \\
&\qquad\qquad y_{ijk} \geq 0, \quad \text{for all } i, \, j, \, k.
\end{aligned}
\right.
\tag{9.2}
$$

A two-level mathematical program is formulated by replacing the outer program of Equation (9.2) from minimize to maximize

$$
\left\{
\begin{aligned}
&Z_v^{\mathcal{L}} = \text{minimize}_{(p,\mathcal{A},\mathcal{B},\mathcal{E}) \in s} \ \text{minimize}_y \ \mathcal{Z} \\
&\qquad = \sum_{i=1}^{m} \sum_{j=1}^{n} p_{ij} y_{ij} + \frac{1}{2} \sum_{i=1}^{m} \sum_{j=1}^{n} \sum_{k=1}^{l} v_{ijk} y_{ijk} \\
&\text{subject to} \\
&\qquad \sum_{j=1}^{n} \sum_{k=1}^{l} y_{ijk} = \mathcal{A}_i, \quad \text{for} \quad i = 1, 2, \ldots, m \\
&\qquad \sum_{i=1}^{m} \sum_{k=1}^{l} y_{ijk} = \mathcal{B}_j, \quad \text{for} \quad j = 1, 2, \ldots, n \\
&\qquad \sum_{i=1}^{m} \sum_{j=1}^{n} y_{ijk} = \mathcal{E}_k, \quad \text{for} \quad k = 1, 2, \ldots, l \\
&\qquad\qquad y_{ijk} \geq 0, \quad \text{for all } i, \, j, \, k.
\end{aligned}
\right.
\tag{9.3}
$$

9.4.2 THE WEIGHTING PROBLEM OF FMQTP IS

Definition 9.4 *A point y^* is called fuzzy efficient solution of FMQTP problem if $\mathcal{Z}^v(y^*) \leq \mathcal{Z}^v(y)$ with $\mathcal{Z}^v(y^*) < \mathcal{Z}^v(y)$ holds for at least one $v = 1, 2, \ldots, \mathcal{Q}$.*

Model 9.1

Consider the multi-objective weighting quadratic transportation problem:

$$
\left\{
\begin{aligned}
\text{minimize}_y \ & \left(\sum_{r=1}^{k} w^r \left(\sum_{i=1}^{m} \sum_{j=1}^{n} p_{ij}^r y_{ij} + \frac{1}{2} \sum_{i=1}^{m} \sum_{j=1}^{n} \sum_{k=1}^{l} v_{ijk}^r y_{ijk} \right) \right) \\
= & \left(w^1 \sum_{i=1}^{m} \sum_{j=1}^{n} p_{ij}^1 y_{ij} + \frac{1}{2} \sum_{i=1}^{m} \sum_{j=1}^{n} \sum_{k=1}^{l} v_{ijk}^1 y_{ijk} \right) \\
& + \cdots + \left(w^r \sum_{i=1}^{m} \sum_{j=1}^{n} p_{ij}^k y_{ij} + \frac{1}{2} \sum_{i=1}^{m} \sum_{j=1}^{n} \sum_{k=1}^{l} q_{ijk}^k y_{ijk} \right) \\
\text{subject to} \ & \\
& \sum_{j=1}^{n} \sum_{k=1}^{l} y_{ijk} = \mathscr{A}_i \\
& \sum_{i=1}^{m} \sum_{k=1}^{l} y_{ijk} = \mathscr{B}_j \\
& \sum_{i=1}^{m} \sum_{j=1}^{n} y_{ijk} = \mathscr{E}_k \text{ and } \mathscr{M} = y \in \mathscr{R}^n, \ y \geq 0
\end{aligned}
\right.
\tag{9.4}
$$

where $w = \left\{ w^r \in \Omega : w^r \geq 0, \ \sum_{r=1}^{k} w^r = 1 \right\}$ is the fuzzy.

Definition 9.5 *A point $\mathscr{X}^*(w^*)$ is called fuzzy optimal solution of the problem, Model 9.1 if, for each $y \in \mathscr{M}$*

$$
\sum_{r=1}^{k} w^r \left(\sum_{i=1}^{m} \sum_{j=1}^{n} p_{ij} y_{ij} + \frac{1}{2} \sum_{i=1}^{m} \sum_{j=1}^{n} \sum_{k=1}^{l} v_{ijk} y_{ijk}^* \right)
$$

$$
\leq \sum_{r=1}^{k} w^r \left(\sum_{i=1}^{m} \sum_{j=1}^{n} p_{ij} y_{ij} + \frac{1}{2} \sum_{i=1}^{m} \sum_{j=1}^{n} \sum_{k=1}^{l} v_{ijk} y_{ijk} \right).
$$

Theorem 9.1 *Let a, $b \in \mathscr{F}_0(\mathscr{R})$, then, for any $\beta \in [0,1]$ the β-level cuts of v_{ijk}, $\mathscr{A}_i$, $\mathscr{B}_j$, $\mathscr{E}_k$, p_{ijk}, $y_{ijk} \in \mathscr{F}\mathscr{R}(\Omega)$ in random interval are denoted by*

$$
\begin{aligned}
(v_{ijk}^r)_\beta &= [(v_{ijk}^r)_\beta^{\mathscr{L}}, (v_{ijk}^r)_\beta^{\mathscr{U}}] \\
(\mathscr{A}_i)_\beta &= [(\mathscr{A}_i)_\beta^{\mathscr{L}}, (\mathscr{A}_i)_\beta^{\mathscr{U}}] \\
(\mathscr{B}_j)_\beta &= [(\mathscr{B}_j)_\beta^{\mathscr{L}}, (\mathscr{B}_j)_\beta^{\mathscr{U}}] \\
(\mathscr{E}_{ij})_\beta &= [(\mathscr{E}_{ij})_\beta^{\mathscr{L}}, (\mathscr{E}_{ij})_\beta^{\mathscr{U}}] \\
(y_{ijk})_\beta &= [(y_{ijk})_\beta^{\mathscr{L}}, (y_{ijk})_\beta^{\mathscr{U}}].
\end{aligned}
$$

Theorem 9.2 *Suppose that $\mathscr{A} \geq 0$, $\mathscr{B} \geq 0$, $\mathscr{E} \geq 0$, $v^r \geq 0$ and $\mathscr{C}^r \geq 0$, $r = 1, 2, \ldots, k$. If $\mathscr{X}^*$ is a fuzzy pseudo random optimal solution of Model 9.1, then for any $\beta \in (0, 1]$, we have*

$$
\begin{cases}
\mathscr{Z}_{\beta}^{\mathscr{L}} = minimize \sum_{r=1}^{k} w^r \left(\sum_{i=1}^{m} \sum_{j=1}^{n} (p_{ij})_{\beta}^{\mathscr{L}} y_{ij} + \frac{1}{2} \sum_{i=1}^{m} \sum_{j=1}^{n} \sum_{k=1}^{l} (v_{ijk})_{\beta}^{\mathscr{L}} y_{ijk} \right) \\[3mm]
\mathscr{Z}_{\beta}^{\mathscr{U}} = minimize \sum_{r=1}^{k} w^r \left(\sum_{i=1}^{m} \sum_{j=1}^{n} (p_{ij})_{\beta}^{\mathscr{U}} y_{ij} + \frac{1}{2} \sum_{i=1}^{m} \sum_{j=1}^{n} \sum_{k=1}^{l} (v_{ijk})_{\beta}^{\mathscr{U}} y_{ijk} \right)
\end{cases}
$$

where $Z = minimize \sum_{r=1}^{k} w^r \left(\sum_{i=1}^{m} \sum_{j=1}^{n} (p_{ij}) y_{ij} + \frac{1}{2} \sum_{i=1}^{m} \sum_{j=1}^{n} \sum_{k=1}^{l} v_{ijk} y_{ijk} \right).$

Proof. Suppose that y^* *is a solution of Model 9.1, then* $y^* \in \mathscr{M}$ *and*

$$
\sum_{r=1}^{k} w^r \left(\sum_{i=1}^{m} \sum_{j=1}^{n} (p_{ij}) y_{ij}^* + \frac{1}{2} \sum_{i=1}^{m} \sum_{j=1}^{n} \sum_{k=1}^{l} v_{ijk} y_{ijk}^* \right)
$$

$$
\leq \sum_{r=1}^{k} w^r \left(\sum_{i=1}^{m} \sum_{j=1}^{n} (p_{ij}) y_{ij} + \frac{1}{2} \sum_{i=1}^{m} \sum_{j=1}^{n} \sum_{k=1}^{l} v_{ijk} y_{ijk} \right). \tag{9.5}
$$

L.H.S of Equation (9.5) becomes

$$
\sum_{r=1}^{k} w^r \left(\sum_{i=1}^{m} \sum_{j=1}^{n} (p_{ij})^r y_{ij}^* + \frac{1}{2} \sum_{i=1}^{m} \sum_{j=1}^{n} \sum_{k=1}^{l} v_{ijk} y_{ijk}^* \right)
$$

$$
= \sum_{r=1}^{k} w^r \left(\sum_{i=1}^{m} \sum_{j=1}^{n} [(p_{ij}^r)_{\beta}^{\mathscr{L}}, (p_{ij}^r)_{\beta}^{\mathscr{U}}][(y_{ij}^*)_{\beta}^{\mathscr{L}}, (y_{ij}^*)_{\beta}^{\mathscr{U}}] \right)
$$

$$
+ \sum_{r=1}^{k} w^r \left(\frac{1}{2} \sum_{i=1}^{m} \sum_{j=1}^{n} \sum_{k=1}^{l} [(v_{ijk}^r)_{\beta}^{\mathscr{L}}, (v_{ijk}^r)_{\beta}^{\mathscr{U}}][(y_{ijk}^*)_{\beta}^{\mathscr{L}}, (y_{ijk}^*)_{\beta}^{\mathscr{U}}] \right)
$$

$$
= \sum_{r=1}^{k} w^r \left(\sum_{i=1}^{m} \sum_{j=1}^{n} (p_{ij}^r)_{\beta}^{\mathscr{L}}, (y_{ij}^*)_{\beta}^{\mathscr{L}} + \frac{1}{2} \sum_{i=1}^{m} \sum_{j=1}^{n} \sum_{k=1}^{l} (v_{ijk}^r)_{\beta}^{\mathscr{L}} (y_{ijk}^*)_{\beta}^{\mathscr{L}} \right)
$$

$$
+ \left(\sum_{i=1}^{m} \sum_{j=1}^{n} (p_{ij}^r)_{\beta}^{\mathscr{U}}, (y_{ij}^*)_{\beta}^{\mathscr{U}} + \frac{1}{2} \sum_{i=1}^{m} \sum_{j=1}^{n} \sum_{k=1}^{l} (v_{ijk}^r)_{\beta}^{\mathscr{U}} (y_{ijk}^*)_{\beta}^{\mathscr{U}} \right)
$$

$$
= \sum_{r=1}^{k} w^r \left(\frac{1}{2} \sum_{i=1}^{m} \sum_{j=1}^{n} \sum_{k=1}^{l} [(v_{ijk}^r)_{\beta}^{\mathscr{L}}, (v_{ijk}^r)_{\beta}^{\mathscr{U}}][(y_{ijk}^*)_{\beta}^{\mathscr{L}}, (y_{ijk}^*)_{\beta}^{\mathscr{U}}] \right)
$$

$$
+ \sum_{r=1}^{k} w^r \left(\sum_{i=1}^{m} \sum_{j=1}^{n} [(p_{ij}^r)_{\beta}^{\mathscr{L}}, (p_{ij}^r)_{\beta}^{\mathscr{U}}][(y_{ij}^*)_{\beta}^{\mathscr{L}}, (y_{ij}^*)_{\beta}^{\mathscr{U}}] \right)
$$

$$
= \sum_{r=1}^{k} w^r \left(\frac{1}{2} \sum_{i=1}^{m} \sum_{j=1}^{n} \sum_{k=1}^{l} [(v_{ijk}^{r\mathscr{L}})_{\beta}, (v_{ijk}^{r\mathscr{U}})_{\beta}][(y_{ijk}^*)_{\beta}^{\mathscr{L}}, (y_{ijk}^*)_{\beta}^{\mathscr{U}}] \right)
$$

$$
+ \sum_{r=1}^{k} w^r \left(\sum_{i=1}^{m} \sum_{j=1}^{n} [(p_{ij}^{r\mathscr{L}})_{\beta}, (p_{ij}^{r\mathscr{U}})_{\beta}][(y_{ij}^*)_{\beta}^{\mathscr{L}}, (y_{ij}^*)_{\beta}^{\mathscr{U}}] \right)
$$

$$
= \sum_{r=1}^{k} w^r \left(\sum_{i=1}^{m} \sum_{j=1}^{n} p_{ij}^r y_{ij}^* + \sum_{i=1}^{m} \sum_{j=1}^{n} \sum_{k=1}^{l} v_{ijk} y_{ijk}^* \right). \tag{9.6}
$$

R.H.S of Equation (9.5) becomes

$$\sum_{r=1}^{k} w^{r} \left(\sum_{i=1}^{m} \sum_{j=1}^{n} (p_{ij}) y_{ij} + \frac{1}{2} \sum_{i=1}^{m} \sum_{j=1}^{n} \sum_{k=1}^{l} v_{ijk} y_{ijk} \right)$$

$$= \sum_{r=1}^{k} w^{r} \left(\sum_{i=1}^{m} \sum_{j=1}^{n} [(p_{ij}^{r})_{\beta}^{\mathscr{L}}, (p_{ij}^{r})_{\beta}^{\mathscr{U}}][(y_{ij})_{\beta}^{\mathscr{L}}, (y_{ij})_{\beta}^{\mathscr{U}}] \right)$$

$$+ \sum_{r=1}^{k} w^{r} \left(\frac{1}{2} \sum_{i=1}^{m} \sum_{j=1}^{n} \sum_{k=1}^{l} [(v_{ijk}^{r})_{\beta}^{\mathscr{L}}, (v_{ijk}^{r})_{\beta}^{\mathscr{U}}][(y_{ijk})_{\beta}^{\mathscr{L}}, (y_{ijk})_{\beta}^{\mathscr{U}}] \right)$$

$$= \sum_{r=1}^{k} w^{r} \left(\sum_{i=1}^{m} \sum_{j=1}^{n} (p_{ij}^{r})_{\beta}^{\mathscr{L}}, (y_{ij})_{\beta}^{\mathscr{L}} + \frac{1}{2} \sum_{i=1}^{m} \sum_{j=1}^{n} \sum_{k=1}^{l} (v_{ijk}^{r})_{\beta}^{\mathscr{L}} (y_{ijk})_{\beta}^{\mathscr{L}} \right)$$

$$+ \left(\sum_{i=1}^{m} \sum_{j=1}^{n} (p_{ij}^{r})_{\beta}^{\mathscr{U}}, (y_{ij})_{\beta}^{\mathscr{U}} + \frac{1}{2} \sum_{i=1}^{m} \sum_{j=1}^{n} \sum_{k=1}^{l} (v_{ijk}^{r})_{\beta}^{\mathscr{U}} (y_{ijk})_{\beta}^{\mathscr{U}} \right)$$

$$= \sum_{r=1}^{k} w^{r} \left(\frac{1}{2} \sum_{i=1}^{m} \sum_{j=1}^{n} \sum_{k=1}^{l} [(v_{ijk}^{r})_{\beta}^{\mathscr{L}}, (v_{ijk}^{r})_{\beta}^{\mathscr{U}}][(y_{ijk})_{\beta}^{\mathscr{L}}, (y_{ijk})_{\beta}^{\mathscr{U}}] \right)$$

$$+ \sum_{r=1}^{k} w^{r} \left(\sum_{i=1}^{m} \sum_{j=1}^{n} [(p_{ij}^{r})_{\beta}^{\mathscr{L}}, (p_{ij}^{r})_{\beta}^{\mathscr{U}}][(y_{ij})_{\beta}^{\mathscr{L}}, (y_{ij})_{\beta}^{\mathscr{U}}] \right)$$

$$= \sum_{r=1}^{k} w^{r} \left(\frac{1}{2} \sum_{i=1}^{m} \sum_{j=1}^{n} \sum_{k=1}^{l} [(v_{ijk}^{r\mathscr{L}})_{\beta}, (v_{ijk}^{r\mathscr{U}})_{\beta}][(y_{ijk})_{\beta}^{\mathscr{L}}, (y_{ijk})_{\beta}^{\mathscr{U}}] \right)$$

$$+ \sum_{r=1}^{k} w^{r} \left(\sum_{i=1}^{m} \sum_{j=1}^{n} [(p_{ij}^{r\mathscr{L}})_{\beta}, (p_{ij}^{r\mathscr{U}})_{\beta}][(y_{ij})_{\beta}^{\mathscr{L}}, (y_{ij})_{\beta}^{\mathscr{U}}] \right)$$

$$= \sum_{r=1}^{k} w^{r} \left(\sum_{i=1}^{m} \sum_{j=1}^{n} p_{ij}^{r} y_{ij} + \sum_{i=1}^{m} \sum_{j=1}^{n} \sum_{k=1}^{l} v_{ijk} y_{ijk} \right) \tag{9.7}$$

from Equations (9.6) and (9.7) we obtain

$$\begin{cases} \sum_{r=1}^{k} w^{r} \left(\sum_{i=1}^{m} \sum_{j=1}^{n} p_{ij}^{\mathscr{L}} y_{ij}^{\mathscr{L}} + \frac{1}{2} \sum_{i=1}^{m} \sum_{j=1}^{n} \sum_{k=1}^{l} v_{ijk}^{\mathscr{L}} y_{ijk}^{\mathscr{L}} \right) \\ \text{and} \\ \sum_{r=1}^{k} w^{r} \left(\sum_{i=1}^{m} \sum_{j=1}^{n} p_{ij}^{\mathscr{U}} y_{ij}^{\mathscr{U}} + \frac{1}{2} \sum_{i=1}^{m} \sum_{j=1}^{n} \sum_{k=1}^{l} v_{ijk}^{\mathscr{U}} y_{ijk}^{\mathscr{U}} \right) \end{cases} \tag{9.8}$$

Equation (9.8) shows that the part of the theorem is correct too. This completes the proof.
□

The two-level program of Equation (9.2) can be used to compute the upper bound of the objective value of the interval quadratic program of Equation (9.1). However, because the outer program and inner program have opposite orientations for optimal, i.e., one for maximization and another for minimization, the solution to Equation (9.2) is not so simple. The Lagrangian dual problem of Equation (9.1) is to maximize $\Lambda(\eta, \zeta)$ over η, $\zeta \geq 0$, where

$$\Lambda(\eta, \zeta) = \sum_{i=1}^{m} \sum_{j=1}^{n} p_{ij} y_{ij} + \frac{1}{2} \sum_{i=1}^{m} \sum_{j=1}^{n} \sum_{k=1}^{l} v_{ijk} y_{ijk}$$

$$+ \sum_{i=1}^{m} \eta_{i} \left(\sum_{j=1}^{n} a_{ij} y_{j} - b_{i} \right) - \sum_{j=1}^{n} \zeta_{j} y_{j}. \tag{9.9}$$

Since in Equation (9.1) is symmetric and positive semi-definite, for a given λ and δ, the function

$$\sum_{i=1}^{m}\sum_{j=1}^{n} p_{ij}y_{ij} + \frac{1}{2}\sum_{i=1}^{m}\sum_{j=1}^{n}\sum_{k=1}^{l} v_{ijk}y_{ijk} + \sum_{i=1}^{m} \eta_i\left(\sum_{j=1}^{n} a_{ij}y_j - b_i\right) - \sum_{j=1}^{n} \zeta_j y_j$$

is convex.

A necessary and sufficient condition for a minimum is that the gradient must vanish, that is

$$\sum_{i=1}^{m} p_{ij}y_i + \sum_{i=1}^{m}\sum_{k=1}^{l} v_{ijk}y_{ik} + \sum_{i=1}^{m} \eta_i(a_{ij}) - \zeta_j = 0, \quad j = 1,2,\ldots,n. \tag{9.10}$$

Thus, the dual form of the inner program in Equation (9.2) can be written as follows:

$$\left\{ \begin{aligned}
&\text{maximize}_{(y,\eta,\zeta)}\,\mathscr{L} = \sum_{i=1}^{m}\sum_{j=1}^{n} p_{ij}y_{ij} + \frac{1}{2}\sum_{i=1}^{m}\sum_{j=1}^{n}\sum_{k=1}^{l} v_{ijk}y_{ijk} \\
&\qquad\qquad\qquad\quad + \sum_{i=1}^{m} \eta_i\left(\sum_{j=1}^{n} a_{ij}y_j - b_i\right) - \sum_{j=1}^{n} \zeta_j y_j \\
&\text{subject to} \\
&\sum_{i=1}^{m} p_{ij}y_i + \sum_{i=1}^{m}\sum_{k=1}^{l} v_{ijk}y_{ik} + \sum_{i=1}^{m} \lambda_i(a_{ij}) - \zeta_j = 0 \\
&\sum_{j=1}^{n}\sum_{k=1}^{l} y_{ijk} = \mathscr{A}_i, \quad \text{for } i = 1,2,\ldots,m \\
&\sum_{i=1}^{m}\sum_{k=1}^{l} y_{ijk} = \mathscr{B}_j, \quad \text{for } j = 1,2,\ldots,n \\
&\sum_{i=1}^{m}\sum_{j=1}^{n} y_{ijk} = \mathscr{E}_k, \quad \text{for } k = 1,2,\ldots,l \\
&\qquad y_{ijk} \geq 0, \text{ for all } i, j, k \quad (\eta_i, \zeta_j) \geq 0.
\end{aligned} \right. \tag{9.11}$$

Now from Equation (9.10), we have

$$\sum_{j=1}^{n}\sum_{i=1}^{m} p_{ij}y_i + \sum_{i=1}^{m} \eta_i\left(\sum_{j=1}^{n} a_{ij}\right) y_j - \sum_{j=1}^{n} \zeta_j y_j = -\sum_{j=1}^{n}\sum_{i=1}^{m}\sum_{k=1}^{l} v_{ijk}y_{ijk}.$$

Substituting this in Equation (9.11), we have,

$$\left\{ \begin{aligned}
&\text{maximize}_{(y,\eta,\zeta)}\,\mathscr{L} = -\frac{1}{2}\sum_{i=1}^{m}\sum_{j=1}^{n}\sum_{k=1}^{l} v_{ijk}y_{ijk} - \sum_{i=1}^{m} b_i\eta_i \\
&\text{subject to} \\
&\sum_{i=1}^{m}\sum_{k=1}^{l} v_{ijk}y_{ik} + \sum_{i=1}^{m} \eta_i a_{ij} - \zeta_j = -\sum_{i=1}^{m} p_{ij}y_i \\
&\sum_{j=1}^{n}\sum_{k=1}^{l} y_{ijk} = \mathscr{A}_i \\
&\sum_{i=1}^{m}\sum_{k=1}^{l} y_{ijk} = \mathscr{B}_j \\
&\sum_{i=1}^{m}\sum_{j=1}^{n} y_{ijk} = \mathscr{E}_k \\
&\qquad y_{ijk} \geq 0, \text{ for all } i, j, k \\
&\quad (\eta_i, \zeta_j) \geq 0, \quad i = 1,2,\ldots,m, \quad j = 1,2,\ldots,n.
\end{aligned} \right. \tag{9.12}$$

9.5 THE SOLUTION PROCEDURE

The writer has designed the fuzzy solid quadratic transportation issue solution approach in this section. Based on the extension principle, the membership function $\mu_{\hat{z}}$ can be defined as:

$$
\begin{aligned}
\mu_{\hat{z}}(z_q) &= \sup_{(p,a,b,e)} \min \mu_{\hat{p}_{ijk}}(p_{ijk}), \mu_{\hat{\mathscr{A}}_i}(a_i), \mu_{\hat{\mathscr{B}}_j}(b_j), \mu_{\hat{\mathscr{E}}_k}(e_k), \ \forall \ i,j,k \\
&= \mathscr{Z}(p,a,b,e);
\end{aligned}
$$

where $\mathscr{Z}(p,a,b,e)$ is defined in Equation (9.1). The application of the extension principle to $\hat{z}$ may be viewed as the application of this extension principle to the β- cuts of $\hat{z}$. Denote the β- cuts of $\hat{\mathscr{C}}_{ijk}$, $\hat{\mathscr{A}}_i$, $\hat{\mathscr{B}}_j$ and $\hat{\mathscr{E}}_k$, as

$$
\begin{aligned}
(\mathscr{C}_{ijk})_\beta &= p_{ijk} \in \mathscr{S}(\hat{\mathscr{C}}_{ijk})|\mu_{(\mathscr{C}_{ijk})}(p_{ijk}) \geq \beta = [(p_{ijk})_\beta^{\mathscr{L}}, (p_{ijk})_\beta^{U}] \\
(\mathscr{A}_i)_\beta &= a_i \in \mathscr{S}(\hat{\mathscr{A}}_i)|\mu_{(A_i)}(a_i) \geq \beta = [(a_i)_\beta^{\mathscr{L}}, (a_i)_\beta^{\mathscr{U}}] \\
(\mathscr{B}_j)_\beta &= b_j \in \mathscr{S}(\hat{\mathscr{B}}_j)|\mu_{(B_j)}(b_j) \geq \beta = [(b_j)_\beta^{\mathscr{L}}, (b_j)_\beta^{\mathscr{U}}] \\
(\mathscr{E}_k)_\beta &= e_k \in \mathscr{S}(\hat{\mathscr{E}}_k)|\mu_{(E_k)}(e_k) \geq \beta = [(e_k)_\beta^{\mathscr{L}}, (e_k)_\beta^{\mathscr{U}}].
\end{aligned}
$$

These gaps show where the amount of shipping price, supply, demand, and transportation are located at the β level of possibility. Lagrangian duality states that if one issue is unbounded, the other is impossible to address. Furthermore, if both issues are solvable, then neither of them have ideal solutions with an identical goal in mind. Alternatively, Equation (9.12) can be reconstructed as

$$
\left\{
\begin{aligned}
&\mathscr{Z}_\beta^{\mathscr{U}} = \text{maximize}_{(y,\lambda,\zeta)} \ \mathscr{Z} = -\frac{1}{2}\sum_{i=1}^{m}\sum_{j=1}^{n}\sum_{k=1}^{l} v_{ijk}y_{ijk} - \sum_{i=1}^{m} b_i \eta_i \\
&\text{subject to} \\
&\sum_{i=1}^{m}\sum_{k=1}^{l} v_{ijk}y_{ijk} + \sum_{i=1}^{m} \eta_i a_{ij} - \zeta_j = -\sum_{i=1}^{m} p_{ij}y_i \\
&(\mathscr{C}_{ij})_\beta^{\mathscr{L}} \leq p_{ij} \leq (\mathscr{C}_{ij})_\beta^{\mathscr{U}}, \ i,j = 1,2,\ldots,m,n \\
&(\mathscr{A}_i)_\beta^{\mathscr{L}} \leq a_i \leq (\mathscr{A}_i)_\beta^{\mathscr{U}}, \ i = 1,2,\ldots,m \\
&(\mathscr{B}_j)_\beta^{\mathscr{L}} \leq b_j \leq (\mathscr{B}_j)_\beta^{\mathscr{U}}, \ j = 1,2,\ldots,n \\
&(\mathscr{E}_k)_\beta^{\mathscr{L}} \leq e_k \leq (\mathscr{E}_k)_\beta^{\mathscr{U}}, \ k = 1,2,\ldots,l \\
&\eta_i, \ \zeta_j, \ y_{ijk} \geq 0.
\end{aligned}
\right.
\tag{9.13}
$$

The generated target value is a global optimum solution that addresses the quadratic programming issue that this framework describes. The upper bound of the intended values for the quadratic fuzzy transportation issue indicates the optimal solution, $\mathscr{Z}^{\mathscr{U}}$.

9.5.1 LOWER BOUND

The linear programming duality theorem makes it apparent that the dual and primal models possess an identical objective value. The result of the Equation (9.3) is:

$$\begin{cases} \mathscr{Z}_v^{\mathscr{L}} = \text{minimize} \sum_{i=1}^{m} \sum_{j=1}^{n} \mathscr{C}_{ij} y_{ij} + \frac{1}{2} \sum_{i=1}^{m} \sum_{j=1}^{n} \sum_{k=1}^{l} v_{ijk} y_{ijk} \\ \text{subject to} \\ \quad \sum_{j=1}^{n} \sum_{k=1}^{l} y_{ijk} = \mathscr{A}_i \\ \quad \sum_{i=1}^{m} \sum_{k=1}^{l} y_{ijk} = \mathscr{B}_j \\ \quad \sum_{i=1}^{m} \sum_{j=1}^{n} y_{ijk} = \mathscr{E}_k \\ \quad (\mathscr{C}_{ij})_{\beta}^{\mathscr{L}} \leq p_{ij} \leq (\mathscr{C}_{ij})_{\beta}^{\mathscr{U}}, \quad i,\, j = 1,2,\ldots,m,n \\ \quad (\mathscr{A}_i)_{\beta}^{\mathscr{L}} \leq a_i \leq (\mathscr{A}_i)_{\beta}^{\mathscr{U}}, \quad i = 1,2,\ldots,m \\ \quad (\mathscr{B}_j)_{\beta}^{\mathscr{L}} \leq b_j \leq (\mathscr{B}_j)_{\beta}^{\mathscr{U}}, \quad j = 1,2,\ldots,n \\ \quad (\mathscr{E}_k)_{\beta}^{\mathscr{L}} \leq e_k \leq (\mathscr{E}_k)_{\beta}^{\mathscr{U}}, \quad k = 1,2,\ldots,l \\ \quad\quad y_{ijk} \geq 0. \end{cases} \tag{9.14}$$

We have $(\mathscr{C}_{ij})_{\beta}^{\mathscr{L}}\, y_{ij} \leq p_{ij} y_{ij} \leq (\mathscr{C}_{ij})_{\beta}^{\mathscr{U}}\, y_{ij}$, which is in Equation (9.14), the context of determining the objective function's minimal value.

$$\text{minimize} \quad \mathscr{Z} = \sum_{i=1}^{m} \sum_{j=1}^{n} (\mathscr{C}_{ij})_{\beta}^{\mathscr{L}}\, y_{ij} + \frac{1}{2} \sum_{i=1}^{m} \sum_{j=1}^{n} \sum_{k=1}^{l} v_{ijk} y_{ijk}.$$

9.5.2 QUADRATIC INTERVAL STP

The aforementioned quadratic interval solid transportation problem (QISTP) incorporates the following form which leads to

$$\begin{cases} \text{minimize} \quad \mathscr{Z}^v = \sum_{i=1}^{m} \sum_{j=1}^{n} \sum_{k=1}^{l} [p_{\mathscr{L}ijk}^q, p_{\mathscr{R}ijk}^q] + \frac{1}{2} \sum_{i=1}^{m} \sum_{j=1}^{n} \sum_{k=1}^{l} v_{ijk} y_{ijk} \\ \text{subject to} \\ \quad \sum_{j=1}^{n} \sum_{k=1}^{l} y_{ijk} = \mathscr{A}_i \\ \quad \sum_{i=1}^{m} \sum_{k=1}^{l} y_{ijk} = \mathscr{B}_j \\ \quad \sum_{i=1}^{m} \sum_{j=1}^{n} y_{ijk} = \mathscr{E}_k \\ \quad\quad y_{ijk} \geq 0, \quad \text{for all } i,\, j,\, k \end{cases} \tag{9.15}$$

where $\mathscr{A}_i = [a_{Li}, a_{Ri}]$, $i = 1,2,\ldots,m$, $\mathscr{B}_j = [b_{Lj}, b_{Rj}]$, $j = 1,2,\ldots,n$, $\mathscr{E}_k = [e_{Lk}, e_{Rk}]$, $k = 1,2,\ldots,l$ are intervals of possible values for the supplies, demands, and conveyance capacities. By stating the lower and upper bounds, the range of the target values in problem (9.15) is capable of being determined. The lower bound's simpler situation is covered initially. The principle of crisp formulation implies that a fuzzy constraint problem can be seen as an accumulation of crisp constraint problems, which is evidently true for Equation (9.14). It is appropriate to refer to constraints of this type as predicates or sets of values for a fixed subset of variables. Consider the following multi-objective QISTP:

$$
\left\{
\begin{aligned}
&\mathscr{Z}_v^{\mathscr{L}} = \text{minimize} \quad \mathscr{Z} \\
&\qquad = \sum_{i=1}^{m}\sum_{j=1}^{n}\sum_{k=1}^{l}[p_{\mathscr{L}ijk}^v, p_{\mathscr{R}ijk}^v] + \frac{1}{2}\sum_{i=1}^{m}\sum_{j=1}^{n}\sum_{k=1}^{l} v_{ijk}y_{ijk} \\
&\text{subject to} \\
&\quad \sum_{j=1}^{n}\sum_{k=1}^{l} y_{ijk} \le a_{\mathscr{R}i}; \qquad \sum_{j=1}^{n}\sum_{k=1}^{l} y_{ijk} \ge a_{\mathscr{L}i}; \\
&\quad \sum_{i=1}^{m}\sum_{k=1}^{l} y_{ijk} \le b_{\mathscr{R}j}; \qquad \sum_{i=1}^{m}\sum_{k=1}^{l} y_{ijk} \ge b_{\mathscr{L}j}; \\
&\quad \sum_{i=1}^{m}\sum_{j=1}^{n} y_{ijk} \le e_{\mathscr{R}k}; \qquad \sum_{i=1}^{m}\sum_{j=1}^{n} y_{ijk} \ge e_{\mathscr{L}k}; \\
&\qquad \mathscr{C}_{ij}^{\mathscr{L}} \le p_{ij} \le \mathscr{C}_{ij}^{\mathscr{R}}, \quad j=1,2,\ldots,n, \ i=1,2,\ldots,m \\
&\qquad \mathscr{A}_i^{\mathscr{L}} \le a_i \le \mathscr{A}_i^{\mathscr{R}}, \quad i=1,2,\ldots,m \\
&\qquad \mathscr{B}_j^{\mathscr{L}} \le b_j \le \mathscr{B}_j^{\mathscr{R}}, \quad j=1,2,\ldots,n \\
&\qquad \mathscr{E}_k^{\mathscr{L}} \le e_k \le \mathscr{E}_k^{\mathscr{R}}, \quad k=1,2,\ldots,l \\
&\qquad v_{ijk}^{\mathscr{L}} \le v_{ijk} \le v_{ijk}^{\mathscr{R}} \\
&\qquad y_{ijk} \ge 0.
\end{aligned}
\right.
\tag{9.16}
$$

Example 9.1 *Consider that the following quadratic fuzzy transportation problem is of the form*

$$
\left\{
\begin{aligned}
&\text{minimize} \ \ \mathscr{Z} = \ (-10,-6)y_{111} + (-5,-3)y_{222} + (1,2)y_{333} + 4y_{111}^2 \\
&\qquad\qquad\qquad + 4y_{222}^2 + 4y_{333}^2 + 8y_{111}y_{222} + 8y_{222}y_{333} + 8y_{333}y_{111} \\
&\text{subject to} \\
&\qquad (2,4)y_{111} + 6y_{222} + (1,2)y_{333} \le (2,20) \\
&\quad\ (4,6)y_{111} + (-1,-0.5)y_{222} + y_{333} \le (4,6) \\
&\quad (-2,8)y_{111} + (1,4)y_{222} + (3,2)y_{333} \le (1,2) \\
&\qquad \sum_{j=1}^{3}\sum_{k=1}^{2} y_{3jk} = [17,21] \\
&\qquad \sum_{j=1}^{3}\sum_{k=1}^{2} y_{1jk} = [24,8] \\
&\qquad \sum_{j=1}^{3}\sum_{k=1}^{2} y_{2jk} = [18,19] \\
&\qquad \sum_{i=1}^{3}\sum_{k=1}^{2} y_{i2k} = [15,17] \\
&\qquad \sum_{i=1}^{3}\sum_{k=1}^{2} y_{i1k} = [2,4] \\
&\qquad \sum_{i=1}^{3}\sum_{k=1}^{2} y_{i3k} = [10,12] \\
&\quad y_{111}, y_{222}, y_{333} \ge 0.
\end{aligned}
\right.
$$

The upper bound of the objective value $Z_\beta^{\mathcal{U}}$ by (9.13) can be formulated as:

$$
\left\{
\begin{aligned}
\mathcal{Z}_\beta^{\mathcal{U}} \;=\;& \text{maximize } -\tfrac{1}{2}(8y_{111}^2 + 8y_{222}^2 + 8y_{333}^2 + 16y_{111}y_{222} \\
& +16y_{222}y_{333} + 16y_{333}y_{111}) - 2\lambda_1 - 4\lambda_2 - \lambda_3 \\
\text{subject to } & \\
& 8y_{111} - 4y_{222} + y_{333} + r_{111} + r_{211} + r_{311} - \Delta_1 = 6 \\
& -4y_{111} + 2y_{222} - 4y_{333} + r_{121} + r_{221} + r_{321} - \Delta_2 = 3 \\
& 2y_{111} - 4y_{222} + 2y_{333} + r_{123} + r_{223} + r_{323} - \Delta_3 = -2 \\
& -10 \le C_1 \le -6 \\
& -5 \le C_2 \le -3 \\
& 1 \le C_3 \le 2 \\
& 2\lambda_1 \le r_{111} \le 4\lambda_1 \\
& 4\lambda_2 \le r_{211} \le 6\lambda_2 \\
& -2\lambda_3 \le r_{311} \le 8\lambda_3 \\
& 6\lambda_1 = r_{121} \\
& -\lambda_2 \le r_{222} \le 0.5\lambda_2 \\
& \lambda_3 \le r_{223} \le 4\lambda_3 \\
& \lambda_1 \le r_{331} \le 2\lambda_1 \\
& \lambda_2 = r_{332} \\
& 3\lambda_3 \le r_{333} \le 2\lambda_3 \\
& \lambda_1,\lambda_2,\lambda_3,\Delta_1,\Delta_2,\Delta_3 \ge 0.
\end{aligned}
\right.
$$

By using Lingo[27], we obtain the stationary point $Z_\beta^{\mathcal{U}} = -277$, which occurs at

$$\lambda_1 = 2,\; \lambda_2 = 4,\; \lambda_3 = 1,\; r_{311} = 6,\; r_{321} = 3,\; \Delta_1 = \Delta_2 = 0,\; \Delta_3 = 2,$$
$$\text{and } y_{312} = 1,\; y_{112} = 1,\; y_{122} = 7,\; y_{222} = 8,\; y_{232} = 10,\; p_3 = 1.$$

The lower bound of the objective value $\mathcal{Z}_\beta^{\mathcal{L}}$ according to model (4.2) can be formulated as:

$$
\begin{aligned}
Z_\beta^{\mathcal{L}} \;=\;& \text{minimize } -10y_{111} - 5y_{222} + y_{333} + \frac{1}{2}\Big[\, 8y_{111}^2 + 8y_{222}^2 \\
& +\; 8y_{333}^2 + 16y_{111}y_{222} + 16y_{222}y_{333} + 16y_{333}y_{111} \,\Big]
\end{aligned}
$$

$$
\left\{
\begin{aligned}
\text{subject to } & \\
& \sum_{j=1}^{3}\sum_{k=1}^{2} y_{3jk} \le 17; \qquad \sum_{j=1}^{3}\sum_{k=1}^{2} y_{3jk} \ge 21; \\
& \sum_{j=1}^{3}\sum_{k=1}^{2} y_{1jk} \le 24; \qquad \sum_{j=1}^{3}\sum_{k=1}^{2} y_{1jk} \ge 8; \\
& \sum_{j=1}^{3}\sum_{k=1}^{2} y_{2jk} \le 18; \qquad \sum_{j=1}^{3}\sum_{k=1}^{2} y_{2jk} \ge 19; \\
& \sum_{i=1}^{3}\sum_{k=1}^{2} y_{i2k} \le 15; \qquad \sum_{i=1}^{3}\sum_{k=1}^{2} y_{i2k} \ge 17; \\
& \sum_{i=1}^{3}\sum_{k=1}^{3} y_{i1k} \le 2; \qquad \sum_{i=1}^{3}\sum_{k=1}^{2} y_{i1k} \ge 4; \\
& \sum_{i=1}^{3}\sum_{k=1}^{2} y_{i3k} \le 10; \qquad \sum_{i=1}^{3}\sum_{k=1}^{2} y_{i3k} \ge 12; \\
& y_{ijk} \ge 0.
\end{aligned}
\right.
$$

The global optimum solution $\mathcal{Z}_\beta^{\mathcal{L}} = -6.250, y_{111} = 1.250, y_{222} = 0$, and $y_{333} = 0$. By combining these two findings, one can arrive at the conclusion that the objective values of this interval linear program fall between -6.250 and -277.

9.6 CONCLUSION

Transportation models are commonly used in supply chains and logistics to save costs. The solid transportation problem takes into account supply and demand as well as transportation capacity in order to meet transportation demands in a way that is both economical and effective. A multi-objective quadratic transportation problem was examined using the two-level mathematical programming paradigm. The theoretical applications are presented, and the optimal solution to the relative quadratic transportation problem is found. The recommended method may effectively determine the range of the objective values for the interval quadratic programming issue, as demonstrated by a practical example. Future research should concentrate on creating a fractional neural network and goal programming approach for engineering disciplines with numerous possible applications.

Conflict of Interest
The authors declare no conflict of interest.

REFERENCES

1. M. Akram , S. Muhammad Umer Shah, M. Ali Al-Shamiri, and S. A. Edalatpanah, Fractional transportation problem under interval-valued Fermatean fuzzy sets, *AIMS Mathematics*, 7: (2022), 17327–17348.
2. M.A. Abo-Sinna, A bi-level non-linear multi-objective decision making under fuzziness, *Journal of Operational Research Society*, 38: (2001), 484–495.
3. S.R. Arora and A. Khurana, Three dimensional fixed charge bi-criterion indefinite quadratic transportation problem, *Yugoslav Journal of Operational Research*, 14: (2004), 83–97.
4. R.E. Bellman and L.A. Zadeh, Decision making in a fuzzy environment, *Management Sciences*, 17: (1970), 141–164.
5. C.R. Bector, S. Chandra, and J. Dutta, *Principles of Optimization Theory*, Narosa Publishing House, New Delhi (2005).
6. A.K. Bit, M.P. Biswal, and S.S. Alam, An additive fuzzy programming model for multi-objective transportation problem, *Fuzzy Sets and Systems*, 57 : (1993), 313–319.
7. J.J. Buckly and T. Feuring, Evolutionary algorithm solution to fuzzy problems: Fuzzy linear programming, *Fuzzy Sets and Systems*, 109 : (2000), 35–53.
8. S. Chanas, The use of parametric programming in fuzzy linear programming, *Fuzzy Sets and Systems*, 11: (1983), 243–251.
9. S. Chanas, W. Kolodziejczyk, and A. Machaj, A fuzzy approach to the transportation problem, *Fuzzy Sets and Systems*, 13: (1984), 211–221.
10. S. Chanas and D. Kuchta, Fuzzy integer transportation problem, *Fuzzy Sets and Systems*, 98: (1998), 291–298.
11. A. Charnes and W.W. Cooper, The stepping-stone method for explaining linear programming calculation in transportation problem, *Management Science*, 1: (1954), 49–99.
12. K. Dahiya and V. Verma, Capacitated transportation problem with bounds on rim conditions, *European Journal of Operational Research*, 178: (2007), 718–737.
13. M. Delgado, J.L. Verdegay, and M.A. Vila, A general model for fuzzy linear programming, *Fuzzy Sets and Systems*, 29: (1989), 21–29.
14. J.A. Diaz, Finding a complete description of all efficient solutions to a multi-objective transportation problem, *Ekonomicko - Matematicki Obzor*, 15: (1979), 62–73.
15. B. George and M.N. Thapa, *Linear Programming Introduction*, Springer Series in Operations Research and Financial Engineering. Springer-Verlag, New York (1997).
16. S.K. Gupta and D. Dangar, Duality in fuzzy quadratic programming with exponential membership functions, *Fuzzy Information and Engineering*, 4: (2010), 337–346.

17. P. Gupta and M.K. Mehlawat, An algorithm for a fuzzy transportation problem to select a new type of coal for a steel manufacturing unit, *TOP*, 15: (2007), 114–137.

18. F.L. Hitchcock, The distribution of a product from several sources to numerous localities, *Journal of Mathematics and Physics*, 20: (1941), 224–230.

19. K.B. Haley, The solid transportation problem, *Operations Research*, 10 : (1962), 448–463.

20. K.B. Haley, The multi index problem, *Operations Research*, 11: (1963), 368–379.

21. R. Horcik, Solution of a system of linear equation with fuzzy numbers, *Fuzzy Sets and Systems*, 159: (2008), 1788–1810.

22. K. Ida, M. Gen, and Y. Li, Neural networks for solving multi criteria solid transportation problem, *Computers and Industrial Engineering*, 31: (1996), 873–877.

23. M. Jain and P.K. Saksena, Time minimizing transportation problem with fractional bottleneck objective function, *Yugoslav Journal of Operational Research*, 22: (2012), 115–129.

24. A. Kumar and A. Kaur, Application of classical transportation methods to find the fuzzy optimal solution of fuzzy transportation problem, *Fuzzy Information and Engineering*, 1: (2011), 81–99.

25. A. Kumar, A. Kaur, and A. Gupta, Fuzzy linear programming approach to solving fuzzy transportation problems with transhipment, *Journal of Mathematical Modelling and Algorithm*, 10: (2011), 163–180.

26. A. Kumar and A. Kaur, Methods for solving unbalanced fuzzy transportation problems, *Operational Research International Journal*, 12: (2012), 287–316.

27. H.C.W. Lau, T.M. Chan, W.T. Tsui, F.T.S. Chan, G.T.S. Ho, and K.L. Choy, A fuzzy guided multi-objective evolutionary algorithm model for solving transportation problem, *Expert Systems with Applications*, 36: (2009), 8255–8268.

28. W. Li and X. Tian, Numerical solution method for general interval quadratic programming, *Applied Mathematics and Computation*, 202: (2008), 589–595.

29. Y. Li, K. Ida, and M. Gen, Improved genetic algorithm for solving multi-objective solid transportation problem with fuzzy numbers, *Computers and Industrial Engineering*, 33: (1997), 589–592.

30. S.T. Liu and R.T. Wang, A numerical solution method to interval quadratic programming, *Applied Mathematics and Computation*, 189: (2007), 1274–1281.

31. S.T. Liu, Quadratic programming with fuzzy parameters: A membership function approach, *Chaos Solitons and Fractals*, 40: (2009), 237–245.

32. N. Mahdavi-Amiri and S.H. Nasseri, Duality results and a dual simplex method for linear programming problems with trapezoidal fuzzy variables, *Fuzzy Sets and Systems*, 17: (2007), 1961–1978.

33. A.H. Marbini and M. Tavana, An extension of the linear programming method with fuzzy parameters, *International Journal of Mathematics in Operations Research*, 3: (2011), 44–55.

34. A. Muhammada, U. Shah, S. Muhammada, and T. Allahviranloo, A new method to determine the Fermatean fuzzy optimal solution of transportation problems, *Journal of Intelligent & Fuzzy Systems*, 44: (2023), 309–328.

35. H.M. Nehi, H.R. Maleki, and M. Mashinchi, A canonical representation for the solution of fuzzy linear systems and fuzzy linear programming problem, *Journal of Applied Mathematics and Computation*, 20: (2006), 345–354.

36. M. OhEigeartaigh, A fuzzy transportation algorithm, *Fuzzy Sets and Systems*, 8: (1982), 235–243.

37. B.A. Ozkok and F. Tiryaki, A compensatory fuzzy approach to multi-objective linear supplier selection problem with multiple-item, *Expert Systems with Applications*, 38: (2011), 11363–11368.

38. B.B. Pal, B.N. Moitra, and S. Sen, A linear Goal Programming approach to multi-objective fractional programming with interval parameter sets, *International Journal of Mathematics in Operations Research*, 3: (2011), 697–714.

39. G. Patel and J. Tripathy, The solid transportation problem and its variants, *International Journal of Management and Systems*, 5: (1989), 17–36.

40. S. Poonam, S.H. Abbas, and V.K. Gupta, Fuzzy transportation problem of triangular numbers with β-cut and ranking technique, *Journal of Engineering*, 2: (2012), 1162–1164.

41. S. Pramanik and T.K. Roy, Multi objective transportation model with fuzzy parameters: Priority based fuzzy goal programming approach, *Journal of Transportation Systems Engineering and Information Technology*, 8: (2008), 40–48.

42. S. Preetvanti and P.K. Saxena, The multi objective time transportation problem with additional restrictions, *European Journal of Operational Research*, 146: (2003), 460–476.

43. A.P. Punnen and R. Zhang, Quadratic bottleneck problems, *Naval Research Logistics*, 58: (2011), 153–164.

44. L. Sahoo, Transportation problem in fuzzy environment , *RAIRO-Operations Research*, 57: (2023), 145–156.

45. M. Tada and H. Ishii, An integer fuzzy transportation problem, *Computers and Mathematics with Applications*, 31: (1996), 71–87.

46. Z. Qiao, Y. Zhang, and G. Wang, On fuzzy random linear programming, *Fuzzy Sets and Systems*, 65: (1994), 31–49.

47. X.L. Wu and Y.K. Liu, Optimizing fuzzy portfolio selection problems by parametric quadratic programming, *Fuzzy Optimization Decision Making*, 11: (2012), 411–449.

48. H. Wu and K. Zhang, A new accelerating method for global non-convex quadratic optimization with non-convex quadratic constraints, *Applied Mathematics and Computation*, 197: (2008), 810–818.

49. F. Xie and R. Jia, Nonlinear fixed charge transportation problem by minimum cost flow-based genetic algorithm, *Computers and Industrial Engineering*, 63: (2012), 763–778.

50. R.R. Yager, Level set and the extension principle for interval valued fuzzy sets and its application to uncertainty measures, *Information Sciences*, 178: (2008), 3565–3576.

51. L. Yang and L. Liu, Fuzzy fixed charge solid transportation problem and algorithm, *Applied Soft Computing*, 7: (2007), 879–889.

52. L.A. Zadeh, Fuzzy sets, *Information and Control*, 8: (1965), 338–353.

53. L.A. Zadeh, Fuzzy sets as a basis for a theory of possibility, *Fuzzy Sets and Systems*, 1: (1978), 3–28.

54. H.J. Zimmermann, Description and Optimization to fuzzy systems, *International Journal of General Systems*, 2: (1976), 209–215.

55. H.J. Zimmermann, Fuzzy programming and linear programming with several objective functions, *Fuzzy Sets and Systems*, 1: (1978), 1: 45–55.

56. H.J. Zimmermann, *Fuzzy Sets Decision: Making and Expert Systems*, Kluwar Academic Publisher, Boston, MA (1987).

57. H.J. Zimmermann, Fuzzy mathematical programming. In: C.S. Shapiroed (ed.), *Encyclopedia of Artificial Intelligence*. John Wiley & Sons, Inc., Hoboken, NJ, vol. 1, pp. 521–528 (1992).

10 Unleashing the Power of Nature-Inspired Optimization

Kiran Bala and Geeta Arora

10.1 INTRODUCTION

A contemporary trend in solving difficult optimization models is the adoption of nature-inspired meta-heuristic algorithms, specifically those based on swarm intelligence (SI). Swarm intelligence algorithms [1] have recently caught the attention of numerous research experts in related domains and were described as "any attempt to build algorithms or distributed problem-solving devices inspired by the collective behaviour of social insect colonies and other animal societies" by Bonabeau et al. [2]. Only social insects, such as termites, bees, wasps, and other ant species, were the subject of his study. The cooperative, trial-and-error behavior of social insects is where swarm intelligence first arose. Swarm members are said to be engaging, lively, dynamic, and possess very little inherent intellect. It is a field of study that simulates populations of interacting agents, or swarms, which have the capacity for self-organization. A common example of a swarm system is an immune system, an ant colony, or a flock of birds. Swarm intelligence-based algorithms [3] and evolution [4] are two significant categories of population-based algorithms.

With the advancement of technology, it appears that conventional optimization techniques are ineffective and unable to handle some real-world issues. As a result, during the past 20 years, researchers have become increasingly interested in the development and investigation of various optimization techniques. The goal of optimization is to increase the advantages of a mathematical model or function while minimizing its disadvantages. The key to find the solution is to optimize the parameters connected to a mathematical model. By minimizing or maximizing linked functions, optimization is a philosophy of approaches that enables us to enhance the operational output or response of a system. Numerous study areas employ optimization techniques to find answers that maximize or reduce a set of relevant parameters. Finding an ideal or nearly ideal solution with minimal computational effort is the main aim of optimization. The categorization of optimization problems is presented in Figure 10.1.

For diverse types of optimization issues, there are currently a number of biologically inspired optimization methods, including the genetic algorithm (GA) [5], particle swarm optimization (PSO) [6], artificial bee colony algorithm (ABC) [7–9], ant colony algorithm (ACO) [10, 11], and differential evolution (DE) [12]. These have been effectively employed to tackle optimization problems in real life.

ABC and PSO are two existing nature-inspired algorithms that have demonstrated significant success in resolving numerical optimization issues. Nevertheless, no algorithm can attain all the benefits, according to the "no free lunch" theory [13]. But due to their adaptable design and simple implementation, these algorithms have gained increasing interest. The performance of these algorithms has already been shown in certain studies to be superior to other evolutionary algorithms, including GA, ACO, differential evolution (DE), and bacteria foraging optimization (BFO) [14]. The ABC has attracted significant attention and has been widely used in numerous research fields, including function optimization [7], clustering analysis [15,16], image processing [17], vehicle routing problems [18], signal processing [19,20], engineering proposals [21], chaotic systems [22], and more. Nowadays, PSO algorithms are used to solve problems with restrictions, multiple objective

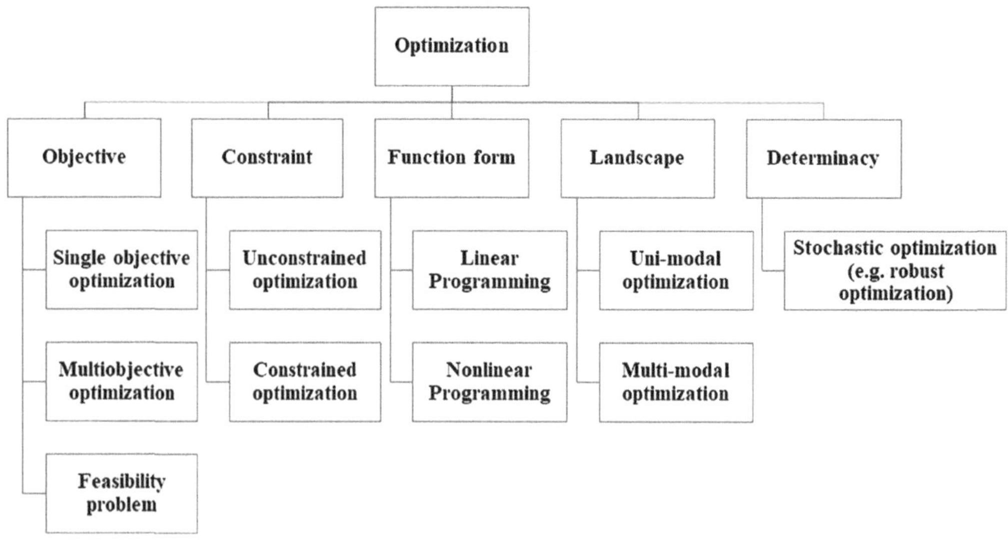

Figure 10.1 Classification of optimization problems.

optimization, dynamically changing environments, and problems with many solutions. PSO has had successful applications in a variety of fields, including function optimization, optimization in cloud computing environments [23], and UAV path planning [24], including telecommunications, system control, data mining, power systems, design, combinatorial optimization, signal processing, network training, and, when used with reinforcement learning, the prediction of pollution, the categorization of plants, the management and navigation of floods, the assessment of water, etc. There are numerous diverse fields in which PSO has applications, including healthcare, the environment, industry, commerce, smart cities, and other general aspects [25, 26].

This work is structured as follows: The original ABC algorithm is explained in Section 10.2. Section 10.3 explains the PSO algorithm with its pseudo-code. In Section 10.4, the conclusion is summarized.

10.2 ARTIFICIAL BEE COLONY (ABC)

The ABC algorithm was originated by Karaboga [27] and is based on swarm intelligence to simulate the foraging behavior of honeybees. ABC offers the benefits of being reliable, efficient, and simple to implement. There are three types of bees in the ABC: employed bees (workers), onlooker bees (observers), and scouts. The best outcome (nectar-rich flower region) is sought after by the algorithm. Bees that were at work, watching, and acting as scouts organized the swarms using the ABC algorithm. The likelihood of other swarms following the one that locates the best food source is higher. Like PSO, ABC keeps track of a user's preferred location. The bee visits a new place and compares it to its existing favorite place before deciding whether to stay or go. If the new position is superior, people will remember it and forget the old one. The memory is still the same as before. The ABC algorithm starts with the deployment of bees and their dispersion in various regions. Bees that are actively searching for nectar or pollen are all employed bees. Working bees convey knowledge about food back to the hive and impart it to interested bees. Bees waiting in the hive for scout bees to report fresh food sources are called observers. Employed bees dance in a designated area to exchange information about food availability. The nectar of the food source affects how the dancing bee dances. Observing bees keep an eye on the dance and decide which food source to use based on its reliability. Employed bees share knowledge with observer bees before returning to the beehive, where they select the most likely to follow. Better food sources therefore draw more bees than inferior ones. All working bees turn into scouts when a food source is depleted. Bees working as scouts and workers follow the exploitation and exploration processes.

Each food source in this method represents a potential resolution to the issue, and the amount of nectar on it represents the fitness value's assessment of its quality. There is exactly one busy bee for every potential food source, and the entire number of food sources equals the total number of bees at work. An observer bee chooses a food source based on the associated probability factor defined below:

$$P_i = \frac{\text{Fit}_i}{\sum_{t=1}^{N_p} \text{Fit}_t} \tag{10.1}$$

Here, the employed bees that are proportional to the ideal solution are evaluating the fitness parameter (objective function value) of the ith solution, and N_p is the number of food sources. Due to this procedure, bees that are employed exchange their information with bees that are watching. The more nectar an employee bee shares, the more likely it is that other bees will choose her. The observer bee uses the following equation to travel to a new spot with the assistance of the selected employed bee:

$$v_i = x_i + \phi_i(x_i - x_j) \tag{10.2}$$

Here, the current position is indicated by x_i, the employed bee selection is indicated by x_j, and ϕ_i is chosen at random from -1 to 1 to locate food sources. After several iterations, every bee who is incapable of discovering a superior food source gets replaced by a scout bee. The scout bees send substitute for the failed bees that hover around any random or unexplored location using the following relation:

$$x_k = L_b + \phi_i(U_b - L_b) \tag{10.3}$$

Here U_b and L_b are the upper and lower bounds, respectively. The process was once again repeated using the employed bees. A position that cannot be improved during the specified cycles is used to determine the parameter limit (number of cycles). The three phases of ABC shown with pseudo-code are defined as follows.

10.2.1 EMPLOYED PHASE

The following considerations should be kept in mind for the production of new solution during this stage:

1. The number of food sources and employed bees is equal.
2. All of the potential solutions have an opportunity.
3. For the creation of a novel solution, an associate is chosen at random.
4. The chosen associate and the present solution shouldn't be the same.
5. For the formulation of a new solution, modifying a randomly chosen variable is essentially given as:

$$x_{\text{new}}^i = x^i + \psi(x^i - x_p^i) \tag{10.4}$$

where $x_p^i = i^{\text{th}}$ variable of pth solution
$x_{\text{new}}^i = i^{\text{th}}$ variable of new solution
ψ is the random variable that lies between -1 and 1
$x^i = i^{\text{th}}$ variable of the current solution

6. Bounds of generated solution.
$x_{\text{new}}^i = l_b$ if $x_{\text{new}}^i < l_b$
$x_{\text{new}}^i = u_b$ if $x_{\text{new}}^i > u_b$

Thus, the objective function value is evaluated, and the fitness of the new solution using the relation is determined after creating it within the boundaries. Choosing the recently generated solution by updating the current solution. Keep track of every solution's trial failures. Reset it if the new solution is better; if not, then increase the trial by one. This updated solution is being considered further. The pseudo-code of this phase is as follows:

1. Enter the initial parameters:
   Objective function, trial, l_b, u_b, $N_p = s/2$ number of food source, here s is swarm size
2. for $k = 1$ to N_p
   Select a partner (p) randomly s.t. $k \neq p$.
   Select a variable (i) randomly and update i^{th} variable $x_{\text{new}}^i = x^i + \psi(x^i - x_p^i)$
3. Bound modified i (x_{new}^i)
4. Evaluate objective function (f_{new}) and fitness (fit_{new})
5. Accept x_{new}, if $\text{fit}_{\text{new}} >$ fitness and set trial $= 0$. Else increase trial by 1.
6. End

10.2.2 ONLOOKER BEES PHASE

There is a condition for entering this phase, which is of the probability of bees exploiting a particular food source. As per the fitness of each food source, calculate their probability, which is determined by Equation (10.1). The Pseudo-code for Onlooker Bees Phase can be defined as follows:

1. Input the parameters: objective function, fit, trial, l_b, u_b, $N_p = s/2$, that denotes the solution to generate with s is the swarm size.
2. Set $m = 0$ and $n = 1$(m corresponds to the food source and n for onlooker bees)
3. While $m < N_p$
   Generate random no. r
   If $r <$ prob
   Randomly select a partner (p) s.t. $n \neq p$.
   Selected a variable (i) randomly and update i^{th} variable $x_{\text{new}}^i = x^i + \psi(x^i - x_p^i)$
   Bound modified i (x_{new}^i)
   Evaluate objective function (fnew) and fitness (fitnew)
   Accept x_{new}, if fitnew of nth solution $>$ fitness of nth solution and Set trial $= 0$. Else increase trial by 1
   $m = m + 1$
   End
   $n = n + 1$
   Modify $n = 1$(if $n > N_p$)
   End.

10.2.3 SCOUT BEES PHASE

As every solution is associated with an individual trial, we need to specify a parameter limit that is a user-specified integer value. For entering the solution in this phase, the value of the trial should be greater than the limit, and the trial of the abandoned solution is reset to zero. Not every solution passes through the scouting phase. Limits can be calculated as $N_p * d$ because of the d-dimensional problem space. This phase can occur only when the trial counter of at least one solution is greater than the limit.

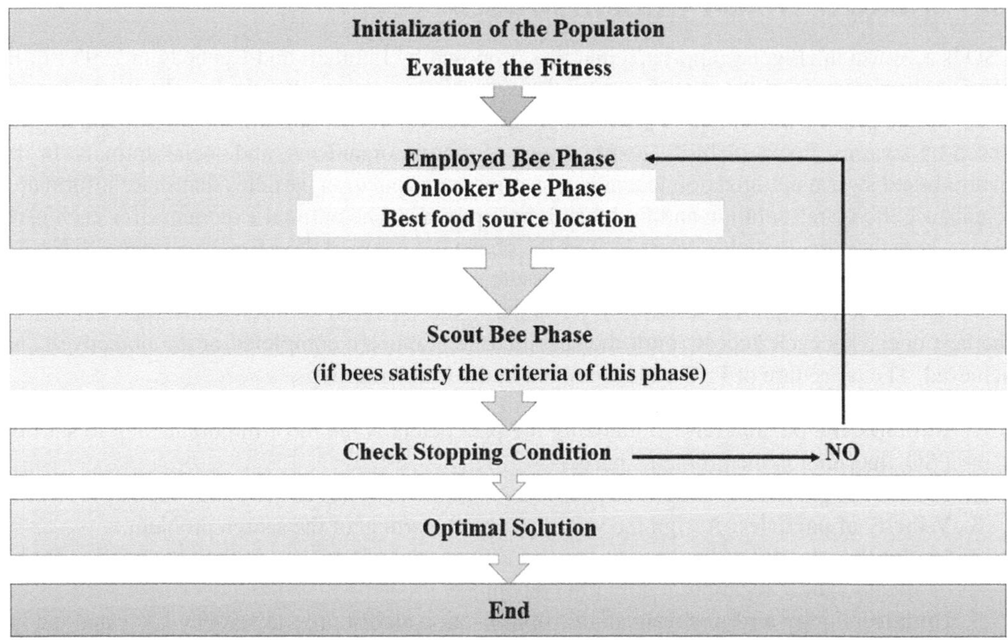

Figure 10.2 Process of ABC algorithm.

Pseudo-Code for Scout Bees Phase Is as Follows:

1. Input = Objective Function, fit, trial, l_b, u_b, limit, P.
2. Identify food source (t) whose trial > limit.
3. Replace X_t with P as $X_t = l_b + \phi_i(u_b - l_b)$
4. Evaluate objective function (f_t) and assign fitness (fit$_t$)

10.2.4 FLOW CHART AND PSEUDO-CODE OF ABC

ABC initializes bee swarms and repeats until stopping criteria are met. ABC optimizes iteratively. Employed and onlooker bees agree on exploitation, while the process of exploration is performed with scout bees.

The all-over process of the basic ABC algorithm is defined in Figure 10.2.

Pseudo-Code of the Algorithm:

1. Input parameters = Population (P), fit, T, l_b, u_b, limit, N_p, Objective function.
2. Calculate objective function value (f) and fitness value (fit).
3. Set trial = 0.
4. for $i = 1$ to T (Iterations)
   Evaluate employed bee stage
   Determined probability
   Apply Onlooker bee phase for generating food source = N_p
   Memorise the best food source
   If trial > limit
   Enter into Scout bee stage
   End
   End

10.3 PARTICLE SWARM ALGORITHM (PSO)

PSO is a swarm intelligence approach that was proposed by Eberhart and Kennedy in 1995 [6]. By altering the pathways of the particles, particle swarm optimization effectively solves optimization issues. Most particle movement is predictable and stochastic. This optimization technique is influenced by swarms, flocks of birds, communities of marine organisms, and social animals. In this nature-based swarm optimization technique, swarms are population particles that share information to enhance the search solution and find the global optimum. The optimal experience for each particle is where they are. Particles' best place on the planet is a result of their best experience. A particle changes its optimal position when it locates a target better than any other. During iterations, each n-particle has a new optimal solution. By comparing all potential solutions, this approach selects the best one. This cycle repeats until the specified iterations are completed or the objective is not achieved. The procedure of PSO can be defined in the following steps:

1. **Initialize the parameters:** Initializing the parameters is the most important step to start the PSO algorithm in the problem space.
2. **Position of particles:** Assign the positions of all the particles in the optimization problem.
3. **Velocity of particles:** Assign the velocity to each particle of the search problem.
4. **Evaluation:** Evaluate the fitness values of each particle. Particle population, which are based on fitness values, are modified by PSO technique.
5. **Update velocity and position:** In this optimization algorithm, the velocity V_i^{k+1} and the position X_i^{k+1} are updated as follows:

$$V_i^{k+1} = V_i^k + a_1 * \text{random} * (P_{\text{best}_i}) - X_i^k) + a_2 * \text{random} * (G_{\text{best}} - X_i^k) \qquad (10.5)$$

$$X_i^{k+1} = X_i^k + V_i^{k+1} \qquad (10.6)$$

   where V_i^k denotes the velocity of particle i at k^{th} iteration, a_1, a_2 are real parameters, random is a random number, whose value lies between 0 and 1, X_i^k is i^{th} is the particle position at k^{th} iteration, P_{best_i} is the personal best position of particle i, G_{best} is the global best position of whole search space.
6. **Repeat step 4**.

10.3.1 FLOW CHART AND PSEUDO-CODE FOR PSO

1. Enter values of parameters fitness, l_b, u_b, limit, N_p, T.
2. Initialization of population (P) randomly and velocity V_i of particle i
3. Calculate objective function (f)
4. Assign P_{best_i} as P and f_{best} as f
5. Evaluate best fitness solution and allocate the solution to G_{best} and fitness to f_{best}
   for $t = 1$ to T
   for $i = 1$ to N_p
   Determine the velocity V_i
   Determine the new position (X_i)
   Bound X_i
   Find objective function value
   Update the population by including X_i and f_i
   Update P_{best} and $f_{p\text{best}}$
   Update G_{best} and $f_{g\text{best}}$
   End
   End

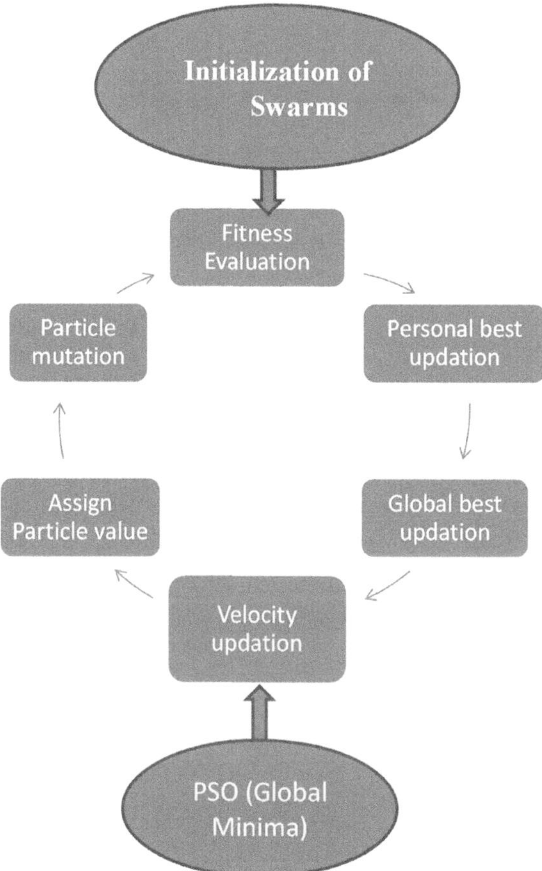

Figure 10.3 Process of general PSO.

PSO is an iterative optimization method for computing that minimizes error by optimizing a problem. It is a statistical process for figuring out parameter values. The particles cooperate, share information, and adhere to a straightforward rule in order to determine the best solution to an optimization problem. It is a novel approach for finding the numerical solutions of nonlinear partial differential equations. It is a technique for improving global search results, and it has a wide range of parameter-space properties. Figure 10.3 presents the original PSO algorithm.

10.4 CONCLUSION

Swarm intelligence-based optimization algorithms are becoming more popular because of their reflectivity and efficient resources. In this chapter, two of these well-liked algorithms, namely, particle swarm optimization (PSO) and artificial bee colony have been examined. Particle swarm optimization (PSO) and the artificial bee colony algorithm (ABC) are two well-known optimization techniques that have been effectively used to solve a variety of optimization issues, particularly in function optimization, and are also presently used for the numerical simulation of various mathematical problems, such as numerical solutions of various partial differential equations, optimizing the shape parameter of the RBF (radial basis function), and many more. These approaches have been used to reduce a number of standard uni-modal and multifunctional models. The overall use of SI-based algorithms (PSO, ABC) for optimization issues was the subject of this review.

REFERENCES

1. Qawqzeh, Y., Gani, A., Siddiqa, A., Ahmedy, I., Hamed, H. N. A., Xia, F., & Kharel, R. (2021). A review of swarm intelligence algorithms deployment for scheduling and optimization in cloud computing environments. *PeerJ Computer Science*, 7, e696.

2. Bonabeau, E., Dorigo, M., & Theraulaz, G. (1999). *Swarm Intelligence: From Natural to Artificial Systems (No. 1)*. Oxford University Press, Oxford.

3. Eberhart, R. C., Shi, Y., & Kennedy, J. (2001). *Swarm Intelligence (Morgan Kaufmann Series in Evolutionary Computation)*. Morgan Kaufmann Publishers, Burlington, MA.

4. Michalewicz, Z., Fogel, D. B., & Beack, T. (2000). *Evolutionary Computation: Vol. 2, Advanced Algorithms and Operators*. Taylor & Francis, New York.

5. Kumar, A. M. S., Parthiban, K., & Shankar, S. S. (2019). An efficient task scheduling in a cloud computing environment using hybrid Genetic Algorithm-Particle Swarm Optimization (GA-PSO) algorithm. *In 2019 International Conference on Intelligent Sustainable Systems (ICISS)*. IEEE. Palladam, India, pp. 29–34.

6. Eberhart, R., & Kennedy, J. (1995). A new optimizer using particle swarm theory. *Proceedings of the Sixth International Symposium on Micro Machine and Human Science (MHS'95)*. IEEE. Nagoya, Japan, pp. 39–43.

7. Chen, C.-F., Chen, C.-Y., Wu, P.-C., & Lin, J.-M. (2020). A new hybrid algorithm based on ABC and PSO for function optimization. *IOP Conference Series: Materials Science and Engineering*, 864(1), 012040.

8. Karaboga, D., & Akay, B. (2009). A comparative study of artificial bee colony algorithm. *Applied Mathematics and Computation*, 214(1), 108–132.

9. Karaboga, D., & Basturk, B. (2007). A powerful and efficient algorithm for numerical function optimization: Artificial bee colony (ABC) algorithm. *Journal of Global Optimization*, 39, 459–471.

10. Dorigo, M., & Stutzle, T. (2004). *Ant Colony Optimization*. MIT Press, Cambridge, MA.

11. Blum, C., Doriga, M. & Stutzle, T. (2005). Ant colony optimization, MIT Press, Cambridge, MA. pp. 300.

12. Zhang, C., Zhou, X., & Zhao, Y. (2009). A novel hybrid differential evolution and particle swarm optimization algorithm for unconstrained optimization. *Operations Research Letters*, 37(2), 117–122.

13. Sarkheyli, A., Zain, A. M., & Sharif, S. (2015). The role of basic, modified and hybrid shuffled frog leaping algorithm on optimization problems: A review. *Soft Computing*, 19, 2011–2038.

14. Passino, K. M. (2002). Biomimicry of bacterial foraging for distributed optimization and control. *IEEE Control Systems Magazine*, 22(3), 52–67.

15. Karaboga, D., & Ozturk, C. (2011). A novel clustering approach: Artificial Bee Colony (ABC) algorithm. *Applied Soft Computing*, 11(1), 652–657.

16. Zhang, C., Ouyang, D., & Ning, J. (2010). An artificial bee colony approach for clustering. *Expert Systems with Applications*, 37(7), 4761–4767.

17. Banharnsakun, A., Achalakul, T., & Sirinaovakul, B. (2011). The best-so-far selection in artificial bee colony algorithm. *Applied Soft Computing*, 11(2), 2888–2901.

18. Szeto, W. Y., Wu, Y., & Ho, S. C. (2011). An artificial bee colony algorithm for the capacitated vehicle routing problem. *European Journal of Operational Research*, 215(1), 126–135.

19. Sabat, S. L., Udgata, S. K., & Abraham, A. (2010). Artificial bee colony algorithm for small signal model parameter extraction of MESFET. *Engineering Applications of Artificial Intelligence*, 23(5), 689–694.

20. Sahin, A. S., Kilic, B., & Kilic, U. (2011). Design and economic optimization of shell and tube heat exchangers using Artificial Bee Colony (ABC) algorithm. *Energy Conversion and Management*, 52(11), 3356–3362.

21. Gao, W.-F., Liu, S.-Y., & Jiang, F. (2011). An improved artificial bee colony algorithm for directing orbits of chaotic systems. *Applied Mathematics and Computation*, 218(7), 3868–3879.

22. Tabrizchi, H., Rafsanjani, M. K., & Balas, V. E. (2021). Multi-task scheduling algorithm based on self-adaptive hybrid ICA–PSO algorithm in cloud environment. *In Soft Computing Applications: Proceedings of the 8th International Workshop Soft Computing Applications (SOFA 2018)*, Vol. II8. Springer International Publishing. Chicago, United States of America.

23. Aslan, S., & Karaboga, D. (2020). A genetic Artificial Bee Colony algorithm for signal reconstruction based big data optimization. *Applied Soft Computing*, 88, 106053.

24. Huang, C., Yang, M., Tan, K. C., & He, L. (2023). Adaptive cylinder vector particle swarm optimization with differential evolution for UAV path planning. *Engineering Applications of Artificial Intelligence*, 121, 105942.

25. Gad, A. G. (2022). Particle swarm optimization algorithm and its applications: A systematic review. *Archives of Computational Methods in Engineering*, 29(5), 2531–2561.

26. Meng, X., Li, H., & Chen, A. (2023). Multi-strategy self-learning particle swarm optimization algorithm based on reinforcement learning. *Mathematical Biosciences and Engineering*, 20(5), 8498–8530.

27. Karaboga, D. (2005). An idea based on honey bee swarm for numerical optimization (Vol. 200). Technical report-tr06, Erciyes University, Engineering Faculty, Computer Engineering Department.

11 EJMD Method for an Unbalanced Fully Intuitionistic Fuzzy Solid Transportation Problem

Shivani and Deepika Rani

11.1 INTRODUCTION

Many real-life applications such as production planning, scheduling, routing, waste management, and inventory control are modeled as a transportation problem which was invented by Hitchcock [10] in 1941. The primary goal of the transportation problem is to deliver the optimal quantity of resources from several sources to different destinations at the minimum possible transportation cost. The transportation problem is also described as a linear programming problem that can be resolved using the simplex approach proposed by Dantzig [6]. Charnes and Cooper [5] developed the stepping stone method to solve the transportation problem. The first step in obtaining an optimal solution for the transportation problems is to identify an initial basic feasible solution [2, 3, 11, 12]. In traditional transportation systems, a single transportation mode is considered for shipping the products from source to destination. However, in real-world scenario, a variety of transportation facilities such as trucks, trains, cargo aircraft, and ships are available with their specializations. Therefore, in a traditional transportation framework when more than one conveyance facility is considered for transportation, the extended problem is known as a solid transportation problem (STP), which was developed by Haley [9] in 1962.

The parameters of the transportation problem are not specified precisely because of the improper information. To tackle this uncertainty in real-life applications, Zadeh [21] developed the fuzzy set (FS) theory. Numerous studies have been conducted by employing FS theory to handle the uncertainty in transportation problems. Kocken and Sivri [13] developed new approach to solve the fuzzy STP in which uncertainty of the parameters are described by triangular fuzzy numbers. Using pentagonal fuzzy numbers, Maheswari and Ganesan [14] examined the uncertainty of fully fuzzy transportation problem. Sadeghi-Moghaddam et al. [19] developed new metaheuristics approaches to solve the transportation problem with fuzzy cost. Ahmad and Adhami [1] investigated the non-linear transportation problem in which objective functions and constraints parameters are considered as parabolic fuzzy numbers. An unbalanced transportation problem involving triangular fuzzy cost coefficient, supply, and demand is solved by Muthuperumal et al. [17].

The membership degree is the only way to represent the uncertainty of the parameters in FS theory. The degree to which an element belongs to the set of feasible solutions may not always be best reflected by the membership function. However, considering the degree of non-membership to examine the presence of hesitating characteristics in a feasible solution set would be preferable. Therefore, Atanassov [4] proposed the intuitionistic fuzzy set (IFS) in which the uncertainty of the parameters is defined in terms of both membership and non-membership degree. Therefore, an IFS is highly beneficial to cope up with unknown quantities. Such real-life problems in which dataset possesses both uncertainty and hesitation can be modeled by intuitionistic fuzzy numbers [16]. Ebrahimnejad and Verdegay [7] developed the new approach to solve the transportation problem with trapezoidal intuitionistic fuzzy parameters. To overcome the shortcomings of existing ap-

DOI: 10.1201/9781003407386-11

Table 11.1
List of Abbreviations

Abbreviations	Full Form
STP	Solid Transportation Problem
FS	Fuzzy Set
IFS	Intuitionistic Fuzzy Set
FIFSTP	Fully Intuitionistic Fuzzy Solid Transportation Problem

---

proaches such as deriving negative solution in the existence of positive cost parameters, Mahmood-irad et al. [15] proposed new effective approach with triangular intuitionistic fuzzy parameters. Ghosh et al. [8] proposed the idea of (α, β)-cut of intuitionistic fuzzy number to convert the intuitionistic fuzzy transportation problem into an interval-valued problem. Rani et al. [18] tackled the uncertainty of non-linear transportation and manufacturing problem with parameters represented as triangular intuitionistic fuzzy numbers.

For the cases in STP in which total supply of sources, total demand of destinations, and total capacity of conveyances all are equal, the problem is balanced STP; otherwise, the problem is an unbalanced STP. In this study, we propose an approach to solve an unbalanced STP in which not only the transportation cost, supply, demand, and capacity of conveyances but also solution variables are considered as triangular intuitionistic fuzzy numbers, and therefore, the problem is named as fully intuitionistic fuzzy solid transportation problem (FIFSTP). In the first stage of proposed solution procedure, the unbalanced FIFSTP is converted into a balanced form by adding either a dummy source/destination/conveyance or all of these. Then, by using the ranking function of triangular intuitionistic fuzzy number, the balanced FIFSTP is converted into a crisp STP. Finally, the LINGO optimization solver is used to find the optimal solution of the proposed problem. All the necessary abbreviations used in this study are included in Table 11.1.

11.2 PRELIMINARIES

Definition 11.1 *[20] The IFS in a universal set Π is defined as $\widetilde{C}^I = \{(\pi, \lambda_{\widetilde{C}^I}(\pi), \gamma_{\widetilde{C}^I}(\pi)) : \pi \in \Pi\}$, where $\lambda_{\widetilde{C}^I}(\pi) : \Pi \to [0,1]$, $\gamma_{\widetilde{C}^I}(\pi) : \Pi \to [0,1]$ denote the membership and non-membership functions, respectively, such that $0 \leq \lambda_{\widetilde{C}^I}(\pi) + \gamma_{\widetilde{C}^I}(\pi) \leq 1, \forall \pi \in \Pi$.*

Definition 11.2 *[8] An IFS, $\widetilde{C}^I = \{(\pi, \lambda_{\widetilde{C}^I}(\pi), \gamma_{\widetilde{C}^I}(\pi)) : \pi \in \Pi\}$ defined on a universal set Π is said to be an intuitionistic fuzzy number if*

(i) $\widetilde{C}^I$ must be a normal, i.e, $\exists \pi \in \Pi$ s.t. $\lambda_{\widetilde{C}^I}(\pi) = 1$.

(ii) $\widetilde{C}^I$ is a convex set, i.e., $\lambda_{\widetilde{C}^I}[\delta \pi_1 + (1-\delta)\pi_2] \geq min\ [\lambda_{\widetilde{C}^I}(\pi_1), \lambda_{\widetilde{C}^I}(\pi_2)], \forall \pi_1, \pi_2 \in \Pi, \delta \in [0,1]$.

(iii) $\widetilde{C}^I$ is a concave set, i.e., $\gamma_{\widetilde{C}^I}[\delta \pi_1 + (1-\delta)\pi_2] \leq max\ [\gamma_{\widetilde{C}^I}(\pi_1), \gamma_{\widetilde{C}^I}(\pi_2)], \forall \pi_1, \pi_2 \in \Pi, \delta \in [0,1]$.

Definition 11.3 *[20] The notations $\widetilde{C}^I = (\sigma^a, \sigma^b, \sigma^c; \widehat{\sigma}^a, \sigma^b, \widehat{\sigma}^c)$ is referred to as a triangular intuitionistic fuzzy number where $\widehat{\sigma}^a \leq \sigma^a \leq \sigma^b \leq \sigma^c \leq \widehat{\sigma}^c$ are real numbers and its membership function $(\lambda_{\widetilde{C}^I}(\pi))$, non-membership function $(\gamma_{\widetilde{C}^I}(\pi))$ are described as:*

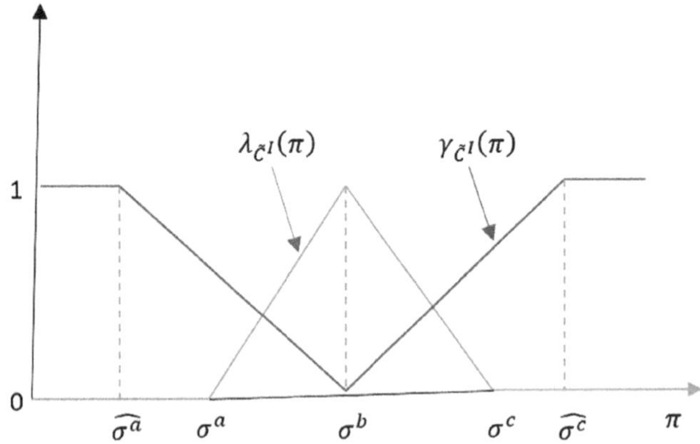

Figure 11.1　Pictorial description of triangular intuitionistic fuzzy number.

$$\lambda_{\widetilde{C}^I}(\pi) = \begin{cases} \frac{\pi - \sigma^a}{\sigma^b - \sigma^a}, & \text{if } \sigma^a < \pi \leq \sigma^b \\ \frac{\sigma^c - \pi}{\sigma^c - \sigma^b}, & \text{if } \sigma^b \leq \pi < \sigma^c \\ 0, & \text{otherwise} \end{cases}$$

$$\gamma_{\widetilde{C}^I}(\pi) = \begin{cases} \frac{\sigma^b - \pi}{\sigma^b - \widehat{\sigma}^a}, & \text{if } \widehat{\sigma}^a < \pi \leq \sigma^b \\ \frac{\pi - \sigma^b}{\widehat{\sigma}^c - \sigma^b}, & \text{if } \sigma^b \leq \pi < \widehat{\sigma}^c \\ 1, & \text{otherwise} \end{cases}$$

These functions are graphically shown in Figure 11.1.

Definition 11.4 *[8] Let $\widetilde{C}^I = (\sigma^a, \sigma^b, \sigma^c; \widehat{\sigma}^a, \sigma^b, \widehat{\sigma}^c)$ and $\widetilde{D}^I = (\eta^a, \eta^b, \eta^c; \widehat{\eta}^a, \eta^b, \widehat{\eta}^c)$ are two triangular intuitionistic fuzzy numbers then $\widetilde{C}^I = \widetilde{D}^I$ if and only if $\widehat{\sigma}^a = \widehat{\eta}^a$, $\sigma^a = \eta^a$, $\sigma^b = \eta^b$, $\sigma^c = \eta^c$, and $\widehat{\sigma}^c = \widehat{\eta}^c$.*

Definition 11.5 *[20] Let $\widetilde{C}^I = (\sigma^a, \sigma^b, \sigma^c; \widehat{\sigma}^a, \sigma^b, \widehat{\sigma}^c)$ and $\widetilde{D}^I = (\eta^a, \eta^b, \eta^c; \widehat{\eta}^a, \eta^b, \widehat{\eta}^c)$ are two triangular intuitionistic fuzzy numbers. Then, arithmetic operations on $\widetilde{C}^I$ and $\widetilde{D}^I$ are defined as*

(i) $\widetilde{C}^I \oplus \widetilde{D}^I = (\sigma^a + \eta^a, \sigma^b + \eta^b, \sigma^c + \eta^c; \widehat{\sigma}^a + \widehat{\eta}^a, \sigma^b + \eta^b, \widehat{\sigma}^c + \widehat{\eta}^c)$

(ii) $\widetilde{C}^I \ominus \widetilde{D}^I = (\sigma^a - \eta^c, \sigma^b - \eta^b, \sigma^c - \eta^a; \widehat{\sigma}^a - \widehat{\eta}^c, \sigma^b - \eta^b, \widehat{\sigma}^c - \widehat{\eta}^a)$

(iii) $k\widetilde{C}^I = \begin{cases} (k\sigma^a, k\sigma^b, k\sigma^c; k\widehat{\sigma}^a, k\sigma^b, k\widehat{\sigma}^c), & \text{if } k \geq 0 \\ (k\sigma^c, k\sigma^b, k\sigma^a; k\widehat{\sigma}^c, k\sigma^b, k\widehat{\sigma}^a), & \text{if } k < 0, \end{cases}$

(iv) $\widetilde{C}^I \otimes \widetilde{D}^I = \Big(\min\{\sigma^a\eta^a, \sigma^a\eta^c, \sigma^c\eta^a, \sigma^c\eta^c\}, \sigma^b\eta^b, \max\{\sigma^a\eta^a, \sigma^a\eta^c, \sigma^c\eta^a, \sigma^c\eta^c\};$

$\min\{\widehat{\sigma}^a\widehat{\eta}^a, \widehat{\sigma}^a\widehat{\eta}^c, \widehat{\sigma}^c\widehat{\eta}^a, \widehat{\sigma}^c\widehat{\eta}^c\}, \sigma^b\eta^b, \max\{\widehat{\sigma}^a\widehat{\eta}^a, \widehat{\sigma}^a\widehat{\eta}^c, \widehat{\sigma}^c\widehat{\eta}^a, \widehat{\sigma}^c\widehat{\eta}^c\} \Big).$

Remark 11.1 *When $\widehat{\sigma}^a$, $\widehat{\eta}^a \geq 0$, $\widetilde{C}^I \otimes \widetilde{D}^I = (\sigma^a\eta^a, \sigma^b\eta^b, \sigma^c\eta^c; \widehat{\sigma}^a\widehat{\eta}^a, \sigma^b\eta^b, \widehat{\sigma}^c\widehat{\eta}^c).$*

Definition 11.6 *[20] Let $\widetilde{C}^I = (\sigma^a, \sigma^b, \sigma^c; \widehat{\sigma}^a, \sigma^b, \widehat{\sigma}^c)$ be a triangular intuitionistic fuzzy number with membership function $\lambda_{\widetilde{C}^I}$ and non-membership function $\gamma_{\widetilde{C}^I}$. The score function of membership and non-membership function are defined as $SC(\lambda_{\widetilde{C}^I}) = \frac{\sigma^a + 2\sigma^b + \sigma^c}{4}$ and $SC(\gamma_{\widetilde{C}^I}) = \frac{\widehat{\sigma}^a + 2\sigma^b + \widehat{\sigma}^c}{4}$, respectively. Then, the accuracy ranking function of triangular intuitionistic fuzzy number $\widetilde{C}^I$ is defined as $\Re(\widetilde{C}^I) = \frac{SC(\lambda_{\widetilde{C}^I}) + SC(\gamma_{\widetilde{C}^I})}{2} = \frac{\widehat{\sigma}^a + \sigma^a + 4\sigma^b + \sigma^c + \widehat{\sigma}^c}{8}.$*

11.3 MATHEMATICAL MODEL OF AN UNBALANCED FULLY INTUITIONISTIC FUZZY SOLID TRANSPORTATION PROBLEM (FIFSTP)

In this section, the model of an unbalanced FIFSTP is proposed in which product is supplied from p sources $(S_l,\ 1 \leq l \leq p)$ to q destinations $(D_m,\ 1 \leq m \leq q)$ through r conveyances $(E_n,\ 1 \leq n \leq r)$. The suggested model makes use of triangular intuitionistic fuzzy numbers to handle the uncertainty of parameters. Each unit of the product being transported has a transportation cost $\widetilde{\chi}^I_{lmn}$ for each route between the source and destination. The amount of product $\widetilde{\mu}^I_l$ is available at each source $(l = 1, 2, \ldots, p)$, amount of product $\widetilde{v}^I_m$ is demanded by each destination $(m = 1, 2, \ldots, q)$, and amount of product $\widetilde{\omega}^I_n$ is the capacity of each conveyance $(n = 1, 2, \ldots, r)$. The aim of this problem is to determine the amount of product $\widetilde{\xi}^I_{lmn}$ that will be conveyed by n^{th} conveyances from the lth source to the mth destination while minimizing the overall transportation cost, which is calculated as below:

11.3.1 MINIMIZE THE TOTAL TRANSPORTATION COST

$$\text{Minimize} \quad \widetilde{\Delta}^I = \underbrace{\sum_{l=1}^{p} \sum_{m=1}^{q} \sum_{n=1}^{r} \widetilde{\chi}^I_{lmn} \widetilde{\xi}^I_{lmn}}_{\text{Transportation cost}} \tag{11.1}$$

The objective function (11.1) minimizes the total transportation cost, which is the sum of the product of the intuitionistic fuzzy per unit transportation cost and the quantity of being transported, across all the sources, destinations, and conveyances.

11.3.2 SUPPLY CONSTRAINTS

The supply constraints ensure that the total amount of product supplied from each source does not exceed the available supply. For the proposed model, in which there are p sources $(S_1, S_2, \ldots, S_p)$, the supply constraints can be written as:

$$\sum_{m=1}^{q} \sum_{n=1}^{r} \widetilde{\xi}^I_{lmn} \leq \widetilde{\mu}^I_l; \quad l = 1, 2, \ldots, p \tag{11.2}$$

11.3.3 DEMAND CONSTRAINTS

The demand constraints ensure that the total quantity received at each destination must meet the specified demand. For the proposed model, in which there are q sources $(D_1, D_2, \ldots, D_q)$, the demand constraints can be written as:

$$\sum_{l=1}^{p} \sum_{n=1}^{r} \widetilde{\xi}^I_{lmn} \geq \widetilde{v}^I_m; \quad m = 1, 2, \ldots, q \tag{11.3}$$

11.3.4 CONVEYANCE CAPACITY CONSTRAINTS

The conveyance capacity constraints ensure that the total quantity transported by each conveyance does not exceed its loading capacity. For the proposed model, in which there are r conveyances $(E_1, E_2, \ldots, E_r)$, the conveyance capacity constraints can be written as:

$$\sum_{l=1}^{p} \sum_{m=1}^{q} \widetilde{\xi}_{lmn}^{I} \leq \widetilde{\omega}_{n}^{I}; \quad n = 1, 2, \ldots, r \tag{11.4}$$

Hence, taking into account the objective function (11.1) and the associated constraints (11.2)–(11.4), the formulated model of FIFSTP is expressed as follows:

Model 11.1

$$\text{Minimize } \widetilde{\Delta}^{I} = \sum_{l=1}^{p} \sum_{m=1}^{q} \sum_{n=1}^{r} \widetilde{\chi}_{lmn}^{I} \otimes \widetilde{\xi}_{lmn}^{I} \tag{11.5}$$

subject to

$$\sum_{m=1}^{q} \sum_{n=1}^{r} \widetilde{\xi}_{lmn}^{I} \leq \widetilde{\mu}_{l}^{I}, \quad l = 1, 2, \ldots, p \tag{11.6}$$

$$\sum_{l=1}^{p} \sum_{n=1}^{r} \widetilde{\xi}_{lmn}^{I} \geq \widetilde{v}_{m}^{I}, \quad m = 1, 2, \ldots, q \tag{11.7}$$

$$\sum_{l=1}^{p} \sum_{m=1}^{q} \widetilde{\xi}_{lmn}^{I} \leq \widetilde{\omega}_{n}^{I}, \quad n = 1, 2, \ldots, r \tag{11.8}$$

The proposed problem (Model 11.1) is assumed to be an unbalanced, i.e., $\sum_{l=1}^{p} \widetilde{\mu}_{l}^{I} \neq \sum_{m=1}^{q} \widetilde{v}_{m}^{I} \neq \sum_{n=1}^{r} \widetilde{\omega}_{n}^{I}$

Remark 11.2 *Supply, demand, and conveyance capacity must be equal to balance the problem, i.e.,* $\sum_{l=1}^{p} \widetilde{\mu}_{l}^{I} = \sum_{m=1}^{q} \widetilde{v}_{m}^{I} = \sum_{n=1}^{r} \widetilde{\omega}_{n}^{I}.$

11.4 PROPOSED APPROACH

To convert an unbalanced FIFSTP into a balanced FIFSTP, a new method called EJMD has been proposed, which has the following steps.

11.4.1 STEP 1: EQUATING TOTAL SUPPLY OF SOURCES TO TOTAL DEMAND OF DESTINATIONS

Let $\sum_{l=1}^{p} \widetilde{\mu}_{l}^{I} = (\mu^{a}, \mu^{b}, \mu^{c}; \widehat{\mu}^{a}, \mu^{b}, \widehat{\mu}^{c})$ and $\sum_{m=1}^{q} \widetilde{v}_{m}^{I} = (v^{a}, v^{b}, v^{c}; \widehat{v}^{a}, v^{b}, \widehat{v}^{c})$

Case 1: If $\widehat{\mu}^{a} \leq \widehat{v}^{a}, \mu^{a} - \widehat{\mu}^{a} \leq v^{a} - \widehat{v}^{a}, \mu^{b} - \mu^{a} \leq v^{b} - v^{a}, \mu^{c} - \mu^{b} \leq v^{c} - v^{b}, \widehat{\mu}^{c} - \mu^{c} \leq \widehat{v}^{c} - v^{c},$ then add a dummy source S_{p+1} with dummy supply $(\mu_{p+1}^{a}, \mu_{p+1}^{b}, \mu_{p+1}^{c}; \widehat{\mu}_{p+1}^{a}, \mu_{p+1}^{b}, \widehat{\mu}_{p+1}^{c})$ where

$$\widehat{\mu}_{p+1}^{a} = \widehat{v}^{a} - \widehat{\mu}^{a}, \ \mu_{p+1}^{a} = v^{a} - \mu^{a}, \ \mu_{p+1}^{b} = v^{b} - \mu^{b}, \ \mu_{p+1}^{c} = v^{c} - \mu^{c}, \ \widehat{\mu}_{p+1}^{c} = \widehat{v}^{c} - \widehat{\mu}^{c}$$

Case 2: If $\widehat{v}^{a} \leq \widehat{\mu}^{a}, v^{a} - \widehat{v}^{a} \leq \mu^{a} - \widehat{\mu}^{a}, v^{b} - v^{a} \leq \mu^{b} - \mu^{a}, v^{c} - v^{b} \leq \mu^{c} - \mu^{b}, \widehat{v}^{c} - v^{c} \leq \widehat{\mu}^{c} - \mu^{c},$ then add a dummy destination D_{q+1} with dummy demand $(v_{q+1}^{a}, v_{q+1}^{b}, v_{q+1}^{c}; \widehat{v}_{q+1}^{a}, v_{q+1}^{b}, \widehat{v}_{q+1}^{c})$ where

$$\widehat{v}_{q+1}^{a} = \widehat{\mu}^{a} - \widehat{v}^{a}, \ v_{q+1}^{a} = \mu^{a} - v^{a}, \ v_{q+1}^{b} = \mu^{b} - v^{b}, \ v_{q+1}^{c} = \mu^{c} - v^{c}, \ \widehat{v}_{q+1}^{c} = \widehat{\mu}^{c} - \widehat{v}^{c}$$

Case 3: If neither Case 1 nor Case 2 are hold, then add a dummy source S_{p+1} with dummy supply $(\mu_{p+1}^a, \mu_{p+1}^b, \mu_{p+1}^c; \widehat{\mu}_{p+1}^a, \mu_{p+1}^b, \widehat{\mu}_{p+1}^c)$ where

$$\widehat{\mu}_{p+1}^a = \max\{0,\ \widehat{v}^a - \widehat{\mu}^a\}$$

$$\mu_{p+1}^a = \widehat{\mu}_{p+1}^a + \max\{0,\ (v^a - \widehat{v}^a) - (\mu^a - \widehat{\mu}^a)\}$$

$$\mu_{p+1}^b = \mu_{p+1}^a + \max\{0,\ (v^b - v^a) - (\mu^b - \mu^a)\}$$

$$\mu_{p+1}^c = \mu_{p+1}^b + \max\{0,\ (v^c - v^b) - (\mu^c - \mu^b)\}$$

$$\widehat{\mu}_{p+1}^c = \mu_{p+1}^c + \max\{0,\ (\widehat{v}^c - v^c) - (\widehat{\mu}^c - \mu^c)\}$$

and add a dummy destination D_{q+1} with dummy demand $(v_{q+1}^a, v_{q+1}^b, v_{q+1}^c; \widehat{v}_{q+1}^a, v_{q+1}^b, \widehat{v}_{q+1}^c)$ where

$$\widehat{v}_{q+1}^a = \max\{0,\ \widehat{\mu}^a - \widehat{v}^a\}$$

$$v_{q+1}^a = \widehat{v}_{q+1}^a + \max\{0,\ (\mu^a - \widehat{\mu}^a) - (v^a - \widehat{v}^a)\}$$

$$v_{q+1}^b = v_{q+1}^a + \max\{0,\ (\mu^b - \mu^a) - (v^b - v^a)\}$$

$$v_{q+1}^c = v_{q+1}^b + \max\{0,\ (\mu^c - \mu^b) - (v^c - v^b)\}$$

$$\widehat{v}_{q+1}^c = v_{q+1}^c + \max\{0,\ (\widehat{\mu}^c - \mu^c) - (\widehat{v}^c - v^c)\}$$

11.4.2 STEP 2: EQUATING TOTAL SUPPLY(DEMAND) OF SOURCES(DESTINATIONS) TO TOTAL CAPACITY OF CONVEYANCES

$$\text{Let } \sum_{l=1}^{s} \widetilde{\mu}_l^I = \sum_{m=1}^{t} \widetilde{v}_m^I = (\tau^a, \tau^b, \tau^c; \widehat{\tau}^a, \tau^b, \widehat{\tau}^c), \text{ where } s = p \text{ or } p+1,\ t = q \text{ or } q+1$$

$$\text{and } \sum_{n=1}^{r} \widetilde{\omega}_n^I = (\omega^a, \omega^b, \omega^c; \widehat{\omega}^a, \omega^b, \widehat{\omega}^c)$$

Case 1: If $\widehat{\tau}^a \leq \widehat{\omega}^a$, $\tau^a - \widehat{\tau}^a \leq \omega^a - \widehat{\omega}^a$, $\tau^b - \tau^a \leq \omega^b - \omega^a$, $\tau^c - \tau^b \leq \omega^c - \omega^b$, $\widehat{\tau}^c - \tau^c \leq \widehat{\omega}^c - \omega^c$, then verify if a dummy source or a dummy destination have been added in Step 1 or not and proceed as per the following subcases:

$$\text{Let } (\psi^a, \psi^b, \psi^c; \widehat{\psi}^a, \psi^b, \widehat{\psi}^c) = (\omega^a - \tau^a, \omega^b - \tau^b, \omega^c - \tau^c; \widehat{\omega}^a - \widehat{\tau}^a, \omega^b - \tau^b, \widehat{\omega}^c - \widehat{\tau}^c)$$

Subcase 1(a): If in Step 1 only a dummy source has been added and no dummy destination has been introduced, then raise the supply $\widetilde{\mu}_{p+1}^I$ of dummy source S_{p+1} by $(\psi^a, \psi^b, \psi^c; \widehat{\psi}^a, \psi^b, \widehat{\psi}^c)$ and also add a dummy destination D_{q+1} with demand $(\psi^a, \psi^b, \psi^c; \widehat{\psi}^a, \psi^b, \widehat{\psi}^c)$.

Subcase 1(b): If in Step 1 only a dummy destination has been introduced and no dummy source has been added, then raise the demand $\widetilde{v}_{q+1}^I$ of dummy destination D_{q+1} by $(\psi^a, \psi^b, \psi^c; \widehat{\psi}^a, \psi^b, \widehat{\psi}^c)$ and also add a dummy source S_{p+1} with supply $(\psi^a, \psi^b, \psi^c; \widehat{\psi}^a, \psi^b, \widehat{\psi}^c)$.

Subcase 1(c): If in Step 1 both a dummy source and a dummy destination have been introduced, then raise the supply $\widetilde{\mu}_{p+1}^I$ of dummy source S_{p+1} and demand $\widetilde{v}_{q+1}^I$ of dummy destination D_{q+1} by $(\psi^a, \psi^b, \psi^c; \widehat{\psi}^a, \psi^b, \widehat{\psi}^c)$.

Subcase 1(d): If in Step 1 neither a dummy source nor a dummy destination has been introduced, then add a dummy source S_{p+1} with supply $(\psi^a, \psi^b, \psi^c; \widehat{\psi}^a, \psi^b, \widehat{\psi}^c)$ and also add a dummy destination D_{q+1} with demand $(\psi^a, \psi^b, \psi^c; \widehat{\psi}^a, \psi^b, \widehat{\psi}^c)$.

Case 2: If $\widehat{\omega}^a \leq \widehat{\tau}^a$, $\omega^a - \widehat{\omega}^a \leq \tau^a - \widehat{\tau}^a$, $\omega^b - \omega^a \leq \tau^b - \tau^a$, $\omega^c - \omega^b \leq \tau^c - \tau^b$, $\widehat{\omega}^c - \omega^c \leq \widehat{\tau}^c - \tau^c$, then add a dummy conveyance E_{r+1} with capacity $(\tau^a - \omega^a, \tau^b - \omega^b, \tau^c - \omega^c; \widehat{\tau}^a - \widehat{\omega}^a, \tau^b - \omega^b, \widehat{\tau}^c - \widehat{\omega}^c)$.

Case 3: If neither Case 1 nor Case 2 are hold, then verify if a dummy source or a dummy destination has been added in Step 1 or not and proceed in accordance with the following subcases. Let

$$\widehat{\phi}^a = \max\{0, \widehat{\omega}^a - \widehat{\tau}^a\}, \; \phi^a = \widehat{\phi}^a + \max\{0, (\omega^a - \widehat{\omega}^a) - (\tau^a - \widehat{\tau}^a)\}$$

$$\phi^b = \phi^a + \max\{0, (\omega^b - \omega^a) - (\tau^b - \tau^a)\}, \; \phi^c = \phi^b + \max\{0, (\omega^c - \omega^b) - (\tau^c - \tau^b)\}$$

$$\widehat{\phi}^c = \phi^c + \max\{0, (\widehat{\omega}^c - \omega^c) - (\widehat{\tau}^c - \tau^c)\}$$

$$\text{and } \widehat{\omega}^a_{r+1} = \max\{0, \widehat{\tau}^a - \widehat{\omega}^a\}, \; \omega^a_{r+1} = \widehat{\omega}^a_{r+1} + \max\{0, (\tau^a - \widehat{\tau}^a) - (\omega^a - \widehat{\omega}^a)\}$$

$$\omega^b_{r+1} = \omega^a_{r+1} + \max\{0, (\tau^b - \tau^a) - (\omega^b - \omega^a)\}, \; \omega^c_{r+1} = \omega^b_{r+1} + \max\{0, (\tau^c - \tau^b) - (\omega^c - \omega^b)\}$$

$$\widehat{\omega}^c_{r+1} = \omega^c_{r+1} + \max\{0, (\widehat{\tau}^c - \tau^c) - (\widehat{\omega}^c - \omega^c)\}$$

Subcase 3(a): If in Step 1 only a dummy source has been added and no dummy destination has been introduced, then raise the supply $\widetilde{\mu}^I_{p+1}$ of dummy source S_{p+1} by $(\phi^a, \phi^b, \phi^c; \widehat{\phi}^a, \phi^b, \widehat{\phi}^c)$ and add a dummy destination D_{q+1} with demand $(\phi^a, \phi^b, \phi^c; \widehat{\phi}^a, \phi^b, \widehat{\phi}^c)$. Also, add a dummy conveyance E_{r+1} with capacity $(\omega^a_{r+1}, \omega^b_{r+1}, \omega^c_{r+1}; \widehat{\omega}^a_{r+1}, \omega^b_{r+1}, \widehat{\omega}^c_{r+1})$.

Subcase 3(b): If in Step 1 only a dummy destination has been introduced and no dummy source has been added, then raise the demand $\widetilde{v}^I_{q+1}$ of dummy destination D_{q+1} by $(\phi^a, \phi^b, \phi^c; \widehat{\phi}^a, \phi^b, \widehat{\phi}^c)$ and add a dummy source S_{p+1} with supply $(\phi^a, \phi^b, \phi^c; \widehat{\phi}^a, \phi^b, \widehat{\phi}^c)$. Also, add a dummy conveyance E_{r+1} with capacity $(\omega^a_{r+1}, \omega^b_{r+1}, \omega^c_{r+1}; \widehat{\omega}^a_{r+1}, \omega^b_{r+1}, \widehat{\omega}^c_{r+1})$.

Subcase 3(c): If in Step 1 both a dummy source and a dummy destination have been introduced, then raise the supply $\widetilde{\mu}^I_{p+1}$ of dummy source S_{p+1} and demand $\widetilde{v}^I_{q+1}$ of dummy destination D_{q+1} by $(\phi^a, \phi^b, \phi^c; \widehat{\phi}^a, \phi^b, \widehat{\phi}^c)$. Also, add a dummy conveyance E_{r+1} with capacity $(\omega^a_{r+1}, \omega^b_{r+1}, \omega^c_{r+1}; \widehat{\omega}^a_{r+1}, \omega^b_{r+1}, \widehat{\omega}^c_{r+1})$.

Subcase 3(d): If in Step 1 neither a dummy source nor a dummy destination has been introduced, then add a dummy source S_{p+1} with supply $(\phi^a, \phi^b, \phi^c; \widehat{\phi}^a, \phi^b, \widehat{\phi}^c)$ and a dummy destination D_{q+1} with demand $(\phi^a, \phi^b, \phi^c; \widehat{\phi}^a, \phi^b, \widehat{\phi}^c)$. Also, add a dummy conveyance E_{r+1} with capacity $(\omega^a_{r+1}, \omega^b_{r+1}, \omega^c_{r+1}; \widehat{\omega}^a_{r+1}, \omega^b_{r+1}, \widehat{\omega}^c_{r+1})$.

11.4.3 STEP 3: FORMULATE THE MODEL OF BALANCED PROBLEM

After applying the Step 1 and Step 2, the proposed problem (Model 11.1) becomes balanced problem, i.e.,

$$\sum_{l=1}^{s} \widetilde{\mu}^I_l = \sum_{m=1}^{t} \widetilde{v}^I_m = \sum_{n=1}^{u} \widetilde{\omega}^I_n, \text{ where } s = p \text{ or } p+1, \; t = q \text{ or } q+1, \text{ and } u = r \text{ or } r+1$$

By considering the unit transportation cost corresponding to dummy source/destination/conveyance as zero triangular intuitionistic fuzzy number, the model of balanced FIFSTP is formulated as below:

Model 11.2

$$\text{Minimize } \widetilde{\Delta}^I = \sum_{l=1}^{s} \sum_{m=1}^{t} \sum_{n=1}^{u} (\chi_{lmn}^{a}, \chi_{lmn}^{b}, \chi_{lmn}^{c}; \widehat{\chi}_{lmn}^{a}, \chi_{lmn}^{b}, \widehat{\chi}_{lmn}^{c}) \tag{11.9}$$

$$\otimes (\xi_{lmn}^{a}, \xi_{lmn}^{b}, \xi_{lmn}^{c}; \widehat{\xi}_{lmn}^{a}, \xi_{lmn}^{b}, \widehat{\xi}_{lmn}^{c})$$

subject to

$$\sum_{m=1}^{t} \sum_{n=1}^{u} (\xi_{lmn}^{a}, \xi_{lmn}^{b}, \xi_{lmn}^{c}; \widehat{\xi}_{lmn}^{a}, \xi_{lmn}^{b}, \widehat{\xi}_{lmn}^{c}) = (\mu_{l}^{a}, \mu_{l}^{b}, \mu_{l}^{c}; \widehat{\mu}_{l}^{a}, \mu_{l}^{b}, \widehat{\mu}_{l}^{c}), \quad l = 1, 2, \ldots, s \tag{11.10}$$

$$\sum_{l=1}^{s} \sum_{n=1}^{u} (\xi_{lmn}^{a}, \xi_{lmn}^{b}, \xi_{lmn}^{c}; \widehat{\xi}_{lmn}^{a}, \xi_{lmn}^{b}, \widehat{\xi}_{lmn}^{c}) = (v_{m}^{a}, v_{m}^{b}, v_{m}^{c}; \widehat{v}_{m}^{a}, v_{m}^{b}, \widehat{v}_{m}^{c}), \quad m = 1, 2, \ldots, t \tag{11.11}$$

$$\sum_{l=1}^{s} \sum_{m=1}^{t} (\xi_{lmn}^{a}, \xi_{lmn}^{b}, \xi_{lmn}^{c}; \widehat{\xi}_{lmn}^{a}, \xi_{lmn}^{b}, \widehat{\xi}_{lmn}^{c}) = (\omega_{n}^{a}, \omega_{n}^{b}, \omega_{n}^{c}; \widehat{\omega}_{n}^{a}, \omega_{n}^{b}, \widehat{\omega}_{n}^{c}), \quad n = 1, 2, \ldots, u \tag{11.12}$$

$$(\xi_{lmn}^{a}, \xi_{lmn}^{b}, \xi_{lmn}^{c}; \widehat{\xi}_{lmn}^{a}, \xi_{lmn}^{b}, \widehat{\xi}_{lmn}^{c}) \geq 0, \quad \forall\, l, m, n \tag{11.13}$$

11.4.4 STEP 4: CONVERT THE BALANCED FIFSTP, I.E., MODEL 11.2 INTO AN EQUIVALENT CRISP LINEAR PROGRAMMING PROBLEM

Using the ranking function of triangular intuitionistic fuzzy number, the Model 11.2 is converted into a crisp model as shown below:

Model 11.3

$$\text{Minimize } \Delta = \sum_{l=1}^{s} \sum_{m=1}^{t} \sum_{n=1}^{u} \frac{1}{8} \left[\widehat{\chi}_{lmn}^{a} \widehat{\xi}_{lmn}^{a} + \chi_{lmn}^{a} \xi_{lmn}^{a} + 4\chi_{lmn}^{b} \xi_{lmn}^{b} + \chi_{lmn}^{c} \xi_{lmn}^{c} + \widehat{\chi}_{lmn}^{c} \widehat{\xi}_{lmn}^{c} \right] \tag{11.14}$$

$$\text{subject to} \tag{11.15}$$

$$\sum_{m=1}^{t} \sum_{n=1}^{u} \widehat{\xi}_{lmn}^{a} = \widehat{\mu}_{l}^{a}, \ \sum_{m=1}^{t} \sum_{n=1}^{u} \xi_{lmn}^{a} = \mu_{l}^{a}, \ \sum_{m=1}^{t} \sum_{n=1}^{u} \xi_{lmn}^{b} = \mu_{l}^{b},$$

$$\sum_{m=1}^{t} \sum_{n=1}^{u} \xi_{lmn}^{c} = \mu_{l}^{c}, \ \sum_{m=1}^{t} \sum_{n=1}^{u} \widehat{\xi}_{lmn}^{c} = \widehat{\mu}_{l}^{c}, \ l = 1, 2, \ldots, s$$

$$\tag{11.16}$$

$$\sum_{l=1}^{s} \sum_{n=1}^{u} \widehat{\xi}_{lmn}^{a} = \widehat{v}_{m}^{a}, \ \sum_{l=1}^{s} \sum_{n=1}^{u} \xi_{lmn}^{a} = v_{m}^{a}, \ \sum_{l=1}^{s} \sum_{n=1}^{u} \xi_{lmn}^{b} = v_{m}^{b},$$

$$\tag{11.17}$$

$$\sum_{l=1}^{s} \sum_{n=1}^{u} \xi_{lmn}^{c} = v_{m}^{c}, \ \sum_{l=1}^{s} \sum_{n=1}^{u} \widehat{\xi}_{lmn}^{c} = \widehat{v}_{m}^{c}, \ m = 1, 2, \ldots, t$$

$$\sum_{l=1}^{s} \sum_{m=1}^{t} \widehat{\xi}_{lmn}^{a} = \widehat{\omega}_{n}^{a}, \ \sum_{l=1}^{s} \sum_{m=1}^{t} \xi_{lmn}^{a} = \omega_{n}^{a}, \ \sum_{l=1}^{s} \sum_{m=1}^{t} \xi_{lmn}^{b} = \omega_{n}^{b},$$

$$\tag{11.18}$$

$$\sum_{l=1}^{s} \sum_{m=1}^{t} \xi_{lmn}^{c} = \omega_{n}^{c}, \ \sum_{l=1}^{s} \sum_{m=1}^{t} \widehat{\xi}_{lmn}^{c} = \widehat{\omega}_{n}^{c}, \ n = 1, 2, \ldots, u$$

$$\widehat{\xi}_{lmn}^{a} \geq 0, \ \xi_{lmn}^{a} - \widehat{\xi}_{lmn}^{a} \geq 0, \geq 0, \ \xi_{lmn}^{b} - \xi_{lmn}^{a} \geq 0,$$

$$\xi_{lmn}^{c} - \xi_{lmn}^{b} \geq 0, \ \widehat{\xi}_{lmn}^{c} - \xi_{lmn}^{c} \geq 0, \ \forall\, l, m, n \tag{11.19}$$

Table 11.2
Input Data for the FIFSTP

	D_1		D_2		D_3		Supply
	E_1	E_2	E_1	E_2	E_1	E_2	
S_1	$(4,7,9;\ 3,7,11)$	$(9,10,12;\ 6,10,15)$	$(6,9,11;\ 5,9,14)$	$(8,11,13;\ 6,11,16)$	$(9,11,13;\ 7,11,15)$	$(7,9,12;5,9,15)$	$(24,27,29;21,27,31)$
S_2	$(7,10,12;\ 5,10,15)$	$(11,13,15;\ 9,13,17)$	$(7,8,10;\ 6,8,13)$	$(6,8,11;\ 4,8,13)$	$(10,13,15;\ 8,13,17)$	$(10,12,14;\ 8,12,16)$	$(28,30,33;\ 25,30,35)$
Conveyance capacity	$(29,31,34;\ 26,31,38)$	$(27,30,33;\ 25,30,36)$	$(29,31,34;\ 26,31,38)$	$(27,30,33;\ 25,30,36)$	$(29,31,34;\ 26,31,38)$	$(27,30,33;\ 25,30,36)$	
Demand	$(15,17,20;\ 12,17,22)$		$(19,22,24;\ 15,22,28)$		$(17,19,22;\ 15,19,25)$		

11.4.5 STEP 5

Solve Model 11.3 with the LINGO optimization solver to get the optimal solution.

Remark 11.3 *In the existing studies, the equality of the intuitionistic fuzzy numbers is considered on the basis of ranking function, which means that the two intuitionistic fuzzy numbers are said to be equal when their ranking values are equal. But it is not correct, i.e., the ranking value of two different intuitionistic fuzzy numbers may be same. For example $\widetilde{A}^I = (3,6,17;\ 2,6,26)$ and $\widetilde{B}^I = (4,7,18;\ 2,6,20)$. In this work, we therefore assumed that two intuitionistic fuzzy numbers are equal if and only if their corresponding elements are the same.*

11.5 NUMERICAL EXAMPLE

In this example, the problem is considered as an unbalanced FIFSTP in which two sources S_1, S_2 supply the products to three destinations D_1, D_2, and D_3 through two conveyances E_1 and E_2. The decision-maker wants to find the optimal solution at minimum possible transportation cost. Because of the insufficient or imprecise information of the input data, the transportation cost, demand, supply, and conveyance capacity have been taken as triangular intuitionistic fuzzy numbers, which is shown in Table 11.2.

The step-by-step explanation of the proposed method to solving this problem is as:

Step 1 From Table 11.2, $\sum_{l=1}^{2} \widetilde{\mu}_l^I \neq \sum_{m=1}^{3} \widetilde{v}_m^I$, i.e., the problem is unbalanced. Neither Case 1 nor Case 2 of Step 1 of the proposed method hold to bring total supply and total demand into balance. Therefore, according to Case 3 of Step 1, it is necessary to add a dummy source S_3 with supply $\widetilde{\mu}_3^I = (3,5,8;0,5,13)$ and dummy destination D_4 with demand $\widetilde{v}_4^I = (4,4,4;\ 4,4,4)$ so that $\sum_{l=1}^{3} \widetilde{\mu}_l^I = \sum_{m=1}^{4} \widetilde{v}_m^I$.

Step 2: Now $\sum_{l=1}^{3} \widetilde{\mu}_l^I = \sum_{m=1}^{4} \widetilde{v}_m^I = (60,67,75;\ 51,67,84)$. Also from Table 11.2, $\sum_{n=1}^{2} \widetilde{\omega}_n^I = (56,61,67;\ 51,61,74)$, therefore, $\sum_{l=1}^{3} \widetilde{\mu}_l^I = \sum_{m=1}^{4} \widetilde{v}_m^I \neq \sum_{n=1}^{2} \widetilde{\omega}_n^I$, i.e., the problem is still unbalanced. Neither Case 1 nor Case 2 of Step 2 of the proposed method hold for equating total supply/demand with total conveyance capacity. Therefore, according to Case 3 (Subcase 3(c)) of Step 2, it is necessary to add a dummy conveyance E_3 with capacity $\widetilde{\omega}_3^I = (4,6,8;0,6,10)$ and increase the supply $\widetilde{\mu}_3^I$ of dummy source S_3 and demand $\widetilde{v}_4^I$ of dummy destination D_4 by $(5,5,5;\ 5,5,5)$ so that the problem becomes balanced.

Step 3: According to Step 3 of the proposed method, considering the unit transportation cost associated with dummy source/destination/conveyance be zero triangular intuitionistic fuzzy number, the model of transformed balanced FIFSTP is formulated as follows:

Model 11.4

$$
\begin{aligned}
\text{Minimize } \widetilde{\Delta}^I = \ & (4,7,9;\ 3,7,11)\otimes\widetilde{\xi}_{111}^I \oplus (9,10,12;\ 6,10,15)\otimes\widetilde{\xi}_{112}^I \\
& \oplus(6,9,11;\ 5,9,14)\otimes\widetilde{\xi}_{121}^I \oplus (8,11,13;\ 6,11,16)\otimes\widetilde{\xi}_{122}^I \\
& \oplus(9,11,13;\ 7,11,15)\otimes\widetilde{\xi}_{131}^I + \oplus(7,9,12;\ 5,9,15)\otimes\widetilde{\xi}_{132}^I \\
& \oplus(7,10,12;\ 5,10,15)\otimes\widetilde{\xi}_{211}^I \oplus (11,13,15;\ 9,13,17)\otimes\widetilde{\xi}_{212}^I \\
& \oplus(7,8,10;\ 6,8,13)\otimes\widetilde{\xi}_{221}^I \oplus (6,8,11;\ 4,8,13)\otimes\widetilde{\xi}_{222}^I
\end{aligned}
$$

$$\oplus(10,13,15;\ 8,13,17)\otimes\widetilde{\xi}^{I}_{231}\oplus(10,12,14;\ 8,12,16)\otimes\widetilde{\xi}^{I}_{232}$$

$$\oplus(0,0,0;\ 0,0,0)\otimes(\widetilde{\xi}^{I}_{113}\oplus\widetilde{\xi}^{I}_{123}\oplus\widetilde{\xi}^{I}_{133}\oplus\widetilde{\xi}^{I}_{141}$$

$$\oplus\widetilde{\xi}^{I}_{142}\oplus\widetilde{\xi}^{I}_{143}\oplus\widetilde{\xi}^{I}_{213}\oplus\widetilde{\xi}^{I}_{223}\oplus\widetilde{\xi}^{I}_{233}\oplus\widetilde{\xi}^{I}_{241}\oplus\widetilde{\xi}^{I}_{242}$$

$$\oplus\widetilde{\xi}^{I}_{243}\oplus\widetilde{\xi}^{I}_{311}\oplus\widetilde{\xi}^{I}_{312}\oplus\widetilde{\xi}^{I}_{313}\oplus\widetilde{\xi}^{I}_{321}\oplus\widetilde{\xi}^{I}_{322}\oplus\widetilde{\xi}^{I}_{323}$$

$$\oplus\widetilde{\xi}^{I}_{331}\oplus\widetilde{\xi}^{I}_{332}\oplus\widetilde{\xi}^{I}_{333}\oplus\widetilde{\xi}^{I}_{341}\oplus\widetilde{\xi}^{I}_{342}\oplus\widetilde{\xi}^{I}_{343})$$

subject to

$$\sum_{m=1}^{4}\sum_{n=1}^{3}(\xi^{a}_{1mn},\xi^{b}_{1mn},\xi^{c}_{1mn};\ \widehat{\xi}^{a}_{1mn},\xi^{b}_{1mn},\widehat{\xi}^{c}_{1mn})=(24,27,29;\ 21,27,31)$$

$$\sum_{m=1}^{4}\sum_{n=1}^{3}(\xi^{a}_{2mn},\xi^{b}_{2mn},\xi^{c}_{2mn};\ \widehat{\xi}^{a}_{2mn},\xi^{b}_{2mn},\widehat{\xi}^{c}_{2mn})=(28,30,33;\ 25,30,35)$$

$$\sum_{m=1}^{4}\sum_{n=1}^{3}(\xi^{a}_{3mn},\xi^{b}_{3mn},\xi^{c}_{3mn};\ \widehat{\xi}^{a}_{3mn},\xi^{b}_{3mn},\widehat{\xi}^{c}_{3mn})=(8,10,13;\ 5,10,18)$$

$$\sum_{l=1}^{3}\sum_{n=1}^{3}(\xi^{a}_{l1n},\xi^{b}_{l1n},\xi^{c}_{l1n};\ \widehat{\xi}^{a}_{l1n},\xi^{b}_{l1n},\widehat{\xi}^{c}_{l1n})=(15,17,20;\ 12,17,22)$$

$$\sum_{l=1}^{3}\sum_{n=1}^{3}(\xi^{a}_{l2n},\xi^{b}_{l2n},\xi^{c}_{l2n};\ \widehat{\xi}^{a}_{l2n},\xi^{b}_{l2n},\widehat{\xi}^{c}_{l2n})=(19,22,24;\ 15,22,28)$$

$$\sum_{l=1}^{3}\sum_{n=1}^{3}(\xi^{a}_{l3n},\xi^{b}_{l3n},\xi^{c}_{l3n};\ \widehat{\xi}^{a}_{l3n},\xi^{b}_{l3n},\widehat{\xi}^{c}_{l3n})=(17,19,22;\ 15,19,25)$$

$$\sum_{l=1}^{3}\sum_{n=1}^{3}(\xi^{a}_{l4n},\xi^{b}_{l4n},\xi^{c}_{l4n};\ \widehat{\xi}^{a}_{l4n},\xi^{b}_{l4n},\widehat{\xi}^{c}_{l4n})=(9,9,9;\ 9,9,9)$$

$$\sum_{l=1}^{3}\sum_{m=1}^{4}(\xi^{a}_{lm1},\xi^{b}_{lm1},\xi^{c}_{lm1};\ \widehat{\xi}^{a}_{lm1},\xi^{b}_{lm1},\widehat{\xi}^{c}_{lm1})=(29,31,34;\ 26,31,38)$$

$$\sum_{l=1}^{3}\sum_{m=1}^{4}(\xi^{a}_{lm2},\xi^{b}_{lm2},\xi^{c}_{lm2};\ \widehat{\xi}^{a}_{lm2},\xi^{b}_{lm2},\widehat{\xi}^{c}_{lm2})=(27,30,33;\ 25,30,36)$$

$$\sum_{l=1}^{3}\sum_{m=1}^{4}(\xi^{a}_{lm3},\xi^{b}_{lm3},\xi^{c}_{lm3};\ \widehat{\xi}^{a}_{lm3},\xi^{b}_{lm3},\widehat{\xi}^{c}_{lm3})=(4,6,8;\ 0,6,10)$$

$$(\xi^{a}_{lmn},\xi^{b}_{lmn},\xi^{c}_{lmn};\ \widehat{\xi}^{a}_{lmn},\xi^{b}_{lmn},\widehat{\xi}^{c}_{lmn})\geq 0\ \ \forall\,l,m,n$$

Step 4: Using Step 4 of the proposed method, the balanced FIFSTP (Model 11.4) is transformed into a corresponding crisp model (Model 11.5) as shown below:

Model 11.5

$$\text{Minimize }\Delta=\quad\frac{1}{8}(3\widehat{\xi}^{a}_{111}+4\xi^{a}_{111}+28\xi^{b}_{111}+9\xi^{c}_{111}+11\widehat{\xi}^{c}_{111}$$

$$+6\widehat{\xi}^{a}_{112}+9\xi^{a}_{112}+40\xi^{b}_{112}+12\xi^{c}_{112}+15\widehat{\xi}^{c}_{112}$$

$$+5\widehat{\xi}^{a}_{121}+6\xi^{a}_{121}+36\xi^{b}_{121}+11\xi^{c}_{121}+14\widehat{\xi}^{c}_{121}$$

$$+6\widehat{\xi}_{122}^{a} + 8\xi_{122}^{a} + 44\xi_{122}^{b} + 13\xi_{122}^{c} + 16\widehat{\xi}_{122}^{c}$$
$$+7\widehat{\xi}_{131}^{a} + 9\xi_{131}^{a} + 44\xi_{131}^{b} + 13\xi_{131}^{c} + 15\widehat{\xi}_{131}^{c}$$
$$+5\widehat{\xi}_{132}^{a} + 7\xi_{132}^{a} + 36\xi_{132}^{b} + 12\xi_{132}^{c} + 15\widehat{\xi}_{132}^{c}$$
$$+5\widehat{\xi}_{211}^{a} + 7\xi_{211}^{a} + 40\xi_{211}^{b} + 12\xi_{211}^{c} + 15\widehat{\xi}_{211}^{c}$$
$$+9\widehat{\xi}_{212}^{a} + 11\xi_{212}^{a} + 52\xi_{212}^{b} + 15\xi_{212}^{c} + 17\widehat{\xi}_{212}^{c}$$
$$+6\widehat{\xi}_{221}^{a} + 7\xi_{221}^{a} + 32\xi_{221}^{b} + 10\xi_{221}^{c} + 13\widehat{\xi}_{221}^{c}$$
$$+4\widehat{\xi}_{222}^{a} + 6\xi_{222}^{a} + 32\xi_{222}^{b} + 11\xi_{222}^{c} + 13\widehat{\xi}_{222}^{c}$$
$$+8\widehat{\xi}_{231}^{a} + 10\xi_{231}^{a} + 52\xi_{231}^{b} + 15\xi_{231}^{c} + 17\widehat{\xi}_{231}^{c}$$
$$+8\widehat{\xi}_{232}^{a} + 10\xi_{232}^{a} + 48\xi_{232}^{b} + 14\xi_{232}^{c} + 16\widehat{\xi}_{232}^{c})$$

subject to

$$\sum_{m=1}^{4}\sum_{n=1}^{3}\widehat{\xi}_{1mn}^{a} = 21, \quad \sum_{m=1}^{4}\sum_{n=1}^{3}\xi_{1mn}^{a} = 24, \quad \sum_{m=1}^{4}\sum_{n=1}^{3}\xi_{1mn}^{b} = 27,$$

$$\sum_{m=1}^{4}\sum_{n=1}^{3}\xi_{1mn}^{c} = 29, \quad \sum_{m=1}^{4}\sum_{n=1}^{3}\widehat{\xi}_{1mn}^{c} = 31$$

$$\sum_{m=1}^{4}\sum_{n=1}^{3}\widehat{\xi}_{2mn}^{a} = 25, \quad \sum_{m=1}^{4}\sum_{n=1}^{3}\xi_{2mn}^{a} = 28, \quad \sum_{m=1}^{4}\sum_{n=1}^{3}\xi_{2mn}^{b} = 30, \quad \sum_{m=1}^{4}\sum_{n=1}^{3}\xi_{2mn}^{c} = 33,$$

$$\sum_{m=1}^{4}\sum_{n=1}^{3}\widehat{\xi}_{2mn}^{c} = 35$$

$$\sum_{m=1}^{4}\sum_{n=1}^{3}\widehat{\xi}_{3mn}^{a} = 5, \quad \sum_{m=1}^{4}\sum_{n=1}^{3}\xi_{3mn}^{a} = 8, \quad \sum_{m=1}^{4}\sum_{n=1}^{3}\xi_{3mn}^{b} = 10, \quad \sum_{m=1}^{4}\sum_{n=1}^{3}\xi_{3mn}^{c} = 13,$$

$$\sum_{m=1}^{4}\sum_{n=1}^{3}\widehat{\xi}_{3mn}^{c} = 18$$

$$\sum_{l=1}^{3}\sum_{n=1}^{3}\widehat{\xi}_{l1n}^{a} = 12, \quad \sum_{l=1}^{3}\sum_{n=1}^{3}\xi_{l1n}^{a} = 15, \quad \sum_{l=1}^{3}\sum_{n=1}^{3}\xi_{l1n}^{b} = 17, \quad \sum_{l=1}^{3}\sum_{n=1}^{3}\xi_{l1n}^{c} = 20,$$

$$\sum_{l=1}^{3}\sum_{n=1}^{3}\widehat{\xi}_{l1n}^{c} = 22$$

$$\sum_{l=1}^{3}\sum_{n=1}^{3}\widehat{\xi}_{l2n}^{a} = 15, \quad \sum_{l=1}^{3}\sum_{n=1}^{3}\xi_{l2n}^{a} = 19, \quad \sum_{l=1}^{3}\sum_{n=1}^{3}\xi_{l2n}^{b} = 22, \quad \sum_{l=1}^{3}\sum_{n=1}^{3}\xi_{l2n}^{c} = 24,$$

$$\sum_{l=1}^{3}\sum_{n=1}^{3}\widehat{\xi}_{l2n}^{c} = 28$$

$$\sum_{l=1}^{3}\sum_{n=1}^{3}\widehat{\xi}_{l3n}^{a} = 15, \quad \sum_{l=1}^{3}\sum_{n=1}^{3}\xi_{l3n}^{a} = 17, \quad \sum_{l=1}^{3}\sum_{n=1}^{3}\xi_{l3n}^{b} = 19, \quad \sum_{l=1}^{3}\sum_{n=1}^{3}\xi_{l3n}^{c} = 22,$$

$$\sum_{l=1}^{3}\sum_{n=1}^{3}\widehat{\xi}_{l3n}^{c} = 25$$

$$\sum_{l=1}^{3}\sum_{n=1}^{3}\widehat{\xi}_{l4n}^{a} = 9, \quad \sum_{l=1}^{3}\sum_{n=1}^{3}\xi_{l4n}^{a} = 9, \quad \sum_{l=1}^{3}\sum_{n=1}^{3}\xi_{l4n}^{b} = 9, \quad \sum_{l=1}^{3}\sum_{n=1}^{3}\xi_{l4n}^{c} = 9,$$

$$\sum_{l=1}^{3}\sum_{n=1}^{3}\widehat{\xi}_{l4n}^{c} = 9$$

$$\sum_{l=1}^{3}\sum_{m=1}^{4}\widehat{\xi}^{a}_{lm1} = 26, \ \sum_{l=1}^{3}\sum_{m=1}^{4}\xi^{a}_{lm1} = 29, \ \sum_{l=1}^{3}\sum_{m=1}^{4}\xi^{b}_{lm1} = 31, \ \sum_{l=1}^{3}\sum_{m=1}^{4}\xi^{c}_{lm1} = 34,$$

$$\sum_{l=1}^{3}\sum_{m=1}^{4}\widehat{\xi}^{c}_{lm1} = 38$$

$$\sum_{l=1}^{3}\sum_{m=1}^{4}\widehat{\xi}^{a}_{lm2} = 25, \ \sum_{l=1}^{3}\sum_{m=1}^{4}\xi^{a}_{lm2} = 27, \ \sum_{l=1}^{3}\sum_{m=1}^{4}\xi^{b}_{lm2} = 30, \ \sum_{l=1}^{3}\sum_{m=1}^{4}\xi^{c}_{lm2} = 33,$$

$$\sum_{l=1}^{3}\sum_{m=1}^{4}\widehat{\xi}^{c}_{lm2} = 36$$

$$\sum_{l=1}^{3}\sum_{m=1}^{4}\widehat{\xi}^{a}_{lm3} = 0, \ \sum_{l=1}^{3}\sum_{m=1}^{4}\xi^{a}_{lm3} = 4, \ \sum_{l=1}^{3}\sum_{m=1}^{4}\xi^{b}_{lm3} = 6, \ \sum_{l=1}^{3}\sum_{m=1}^{4}\xi^{c}_{lm3} = 8,$$

$$\sum_{l=1}^{3}\sum_{m=1}^{4}\widehat{\xi}^{c}_{lm3} = 10$$

$$\widehat{\xi}^{a}_{lmn} \geq 0, \ \xi^{a}_{lmn} - \widehat{\xi}^{a}_{lmn} \geq 0, \ \xi^{b}_{lmn} - \xi^{a}_{lmn} \geq 0, \ \xi^{c}_{lmn} - \xi^{b}_{lmn} \geq 0, \ \widehat{\xi}^{c}_{lmn} - \xi^{c}_{lmn} \geq 0 \ \ \forall \, l, m, n.$$

Step 5: On solving the Model 11.5 using LINGO software the optimal solution is obtained as

$$\widetilde{\xi}^{I}_{111} = (14,16,18; \ 12,16,20), \ \widetilde{\xi}^{I}_{132} = (9,9,9; \ 9,9,9), \ \widetilde{\xi}^{I}_{221} = (0,0,1; \ 0,0,1),$$

$$\widetilde{\xi}^{I}_{222} = (15,16,16; \ 15,16,16)\widetilde{\xi}^{I}_{232} = (1,1,1; \ 1,1,1), \ \widetilde{\xi}^{I}_{123} = (1,1,1; \ 0,1,1),$$

$$\widetilde{\xi}^{I}_{133} = (0,1,1; \ 0,1,1), \ \widetilde{\xi}^{I}_{241} = (9,9,9; \ 9,9,9)\widetilde{\xi}^{I}_{213} = (1,1,1; \ 0,1,1),$$

$$\widetilde{\xi}^{I}_{233} = (2,2,3; \ 0,2,5), \ \widetilde{\xi}^{I}_{223} = (0,1,2; \ 0,1,2), \ \widetilde{\xi}^{I}_{331} = (5,5,5; \ 5,5,6)$$

$$\widetilde{\xi}^{I}_{321} = (1,1,1; \ 0,1,2), \ \widetilde{\xi}^{I}_{322} = (2,3,3; \ 0,3,6), \ \widetilde{\xi}^{I}_{332} = (0,1,3; \ 0,1,3),$$

$$\widetilde{\xi}^{I}_{312} = (0,0,1; \ 0,0,1)$$

and at this solution the optimal intuitionistic fuzzy transportation cost is (219, 333, 470; 149, 333, 592).

11.6 CONCLUSION AND FUTURE RESEARCH SCOPE

Intuitionistic fuzzy set theory is practically more applicable than the fuzzy set theory to express uncertainty in optimization problems, since it considers both the degree of membership and non-membership of an element in the set. To examine this advantage, an unbalanced solid transportation problem is investigated in fully intuitionistic fuzzy environment. An existing method called the JMD method for solving the transportation problem in fully intuitionistic fuzzy environment is extended to solve the proposed problem, and the new method is named as EJMD. The main advantage of the EJMD method is that it provides an intuitionistic fuzzy solution to the intuitionistic fuzzy problem rather than the crisp solution. As a result, this approach widens the range of the obtained solution and also effectively handles ambiguity and imprecision.

In future, the proposed method can be extended to solve an unbalanced fully intuitionistic fuzzy multi-objective solid transportation problem. Also, solving an unbalanced fractional or non-linear solid transportation problem in fully intuitionistic fuzzy environment will be an interesting problem.

REFERENCES

1. Ahmad, F., & Adhami, A. Y. (2019). Neutrosophic programming approach to multiobjective nonlinear transportation problem with fuzzy parameters. *International Journal of Management Science and Engineering Management*, 14(3), 218–229.
2. Ahmed, M. M., Khan, A. R., Uddin, M. S., & Ahmed, F. (2016). A new approach to solve transportation problems. *Open Journal of Optimization,* 5(1), 22–30.
3. Amaliah, B., Fatichah, C., & Suryani, E. (2022). A new heuristic method of finding the initial basic feasible solution to solve the transportation problem. *Journal of King Saud University-Computer and Information Sciences*, 34(5), 2298–2307.
4. Atanassov, K. T. (1986). Intuitionistic fuzzy sets. *Fuzzy Sets and Systems*, 20(1), 87–96.
5. Charnes, A., & Cooper, W. W. (1954). The stepping stone method of explaining linear programming calculations in transportation problems. *Management Science*, 1(1), 49–69.
6. Dantzig, G. B. (1951). Application of the simplex method to a transportation problem. *Activity Analysis and Production and Allocation*.
7. Ebrahimnejad, A., & Verdegay, J. L. (2018). A new approach for solving fully intuitionistic fuzzy transportation problems. *Fuzzy Optimization and Decision Making*, 17(4), 447–474.
8. Ghosh, S., Roy, S. K., Ebrahimnejad, A., & Verdegay, J. L. (2021). Multi-objective fully intuitionistic fuzzy fixed-charge solid transportation problem. *Complex & Intelligent Systems*, 7, 1009–1023.
9. Haley, K. B. (1962). New methods in mathematical programming-the solid transportation problem. *Operations Research*, 10(4), 448–463.
10. Hitchcock, F. L. (1941). The distribution of a product from several sources to numerous localities. *Journal of Mathematics and Physics*, 20(4), 224–230.
11. Hussein, H. A., Shiker, M. A., & Zabiba, M. S. (2020). A new revised efficient of VAM to find the initial solution for the transportation problem. *In Journal of Physics: Conference Series*, 1591(1), 012032.
12. Karagul, K., & Sahin, Y. (2020). A novel approximation method to obtain initial basic feasible solution of transportation problem. *Journal of King Saud University-Engineering Sciences*, 32(3), 211–218.
13. Kocken, H. G., & Sivri, M. (2016). A simple parametric method to generate all optimal solutions of fuzzy solid transportation problem. *Applied Mathematical Modelling*, 40(8), 4612–4624.
14. Maheswari, P. U., & Ganesan, K. (2018). Solving fully fuzzy transportation problem using pentagonal fuzzy numbers. *In Journal of Physics: Conference Series*, 1000(1), 012014.
15. Mahmoodirad, A., Allahviranloo, T., & Niroomand, S. (2019). A new effective solution method for fully intuitionistic fuzzy transportation problem. *Soft Computing*, 23(12), 4521–4530.
16. Mishra, A., & Kumar, A. (2020). JMD method for transforming an unbalanced fully intuitionistic fuzzy transportation problem into a balanced fully intuitionistic fuzzy transportation problem. *Soft Computing*, 24, 15639–15654.
17. Muthuperumal, S., Titus, P., & Venkatachalapathy, M. (2020). An algorithmic approach to solve unbalanced triangular fuzzy transportation problems. *Soft Computing*, 24(24), 18689–18698.
18. Rani, D., Ebrahimnejad, A., & Gupta, G. (2022). Generalized techniques for solving intuitionistic fuzzy multi-objective non-linear optimization problems. *Expert Systems with Applications*, 202, 117264.
19. Sadeghi-Moghaddam, S., Hajiaghaei-Keshteli, M., & Mahmoodjanloo, M. (2019). New approaches in metaheuristics to solve the fixed charge transportation problem in a fuzzy environment. *Neural Computing and Applications*, 31(1), 477–497.
20. Singh, S. K., & Yadav, S. P. (2018). Intuitionistic fuzzy multi-objective linear programming problem with various membership functions. *Annals of Operations Research*, 269, 693–707.
21. Zadeh, L. A. (1965). Fuzzy sets. *Information and Control*, 8(3), 338–353.

12 Discrete Fourier Series Using Generalized Difference Operator

D. Brightlin and G. Dominic Babu

12.1 INTRODUCTION

In 1807, Fourier surprised a portion of his counterparts by stating that an inconsistent capability could be communicated as a straight blend of sine and cosine capability. For a brief yet magnificent record of the historical backdrop of this subject and its effect on the improvement of science, these straight mix presently called Fourier series have turned into an imperative device in the examination of specific occasional peculiarities, which are applied in physics and engineering. In 1989, Miller and Rose introduced the discrete analog of the Riemann Libreville partial subsidiary and demonstrated a few properties of the fragmentary contrast administrator. The overall partial distinction Riemann Libreville administrator and its converse $\Delta_V^{-v} f(t)$ were mentioned as application of Δ_V^{-v} by taking $v = m$ and $h = l$, then, at that point, the amount of the mth particle aggregates on nth powers mathematical movement and results of n continuous terms of a number juggling movement have been determined utilizing Jerzy Popenda presented the distinctive administrator Δ_l characterized on $U(k)$ as $\Delta_l U(k) = U(k+1) - U(k)$. In 1989 [8], Miller and Rose presented the discrete simple of the Riemann Libreville partial derivative, demonstrated since properties of the opposite fragmentary difference operator Δ_V^{-v}.

12.2 DISCRETE FOURIER SERIES ARISING FROM GENERALIZED DIFFERENCE OPERATOR

12.2.1 INTRODUCTION

The essential issues in the hypothesis of discrete Fourier series are best depicted in the setting of a more broad discipline known as the hypothesis discrete symmetrical capabilities. Accordingly, we present some phrasing concerning discrete symmetrical capability.

12.2.2 DEFINITION OF DISCRETE FOURIER SERIES

Definition 12.1 *Let $S_l = \{\phi_0, \phi_1, \phi_2, \ldots \phi_M\}$ be an orthonormal on $I = [a, b]$ and assume that u is complex-valued bounded function on $I = [a, b]$. The notation*

$$u(k) \approx \sum_{n=0}^{M} c_n \phi_n(k) \tag{12.1}$$

which means the numbers $c_0, c_1, c_1, \ldots$ are given by the formulas:

$$c_n = (u, \phi_n)_l = l\Delta_l^{-1} (u(k)\phi_n(k)) \Big|_a^b, n = 0, 1, 2, \ldots \tag{12.2}$$

The series in Equation (12.1) is called the Fourier series of u relative to S_l and the numbers $c_0, c_1, c_2, \ldots$ are called the discrete Fourier coefficients of u relative to S_l.

DOI: 10.1201/9781003407386-12

Remark 12.1 *Since* $\{\sin nk\}, \{\cos nk\}$ *are orthonormal system of functions on* $[0, 2\pi]$, *we have discrete Fourier Series as*

$$u(k) = \frac{a_0}{2} + \sum_{v=1}^{V-1} (a_v \cos nk + b_v \sin vk) + \frac{a_V}{2} \cos Vk,$$

where, $a_n = \dfrac{l}{\pi} \Delta_l^{-1} u(k) \cos nk \Big|_0^{2\pi}$, $b_n = \dfrac{l}{\pi} \Delta_l^{-1} u(k) \sin nk \Big|_0^{2\pi}$, *and* $l = \dfrac{\pi}{N}$

12.2.3 INVERSE ON TRIGNOMETRIC FUNCTION

In this section, we derive inverse of the product of general function $u(k)$ with $\sin nk$ & $\cos nk$ and given numerical example of discrete Fourier series.

Theorem 12.1 *If u is real valued bounded and p is an integer, then* $l\Delta_l^{-1} |u(k)| \sin pk \Big|_0^{\infty} = 0$

Proof *For $u(k) \geq 0$. since $u(k)$ is bounded $\Rightarrow |u(k)| \leq m$*

$$l\Delta_l^{-1} |u(k)| \sin pk \Big|_0^{\infty} \leq l\Delta_l^{-1} m \sin pk \Big|_0^{\infty}$$

$$l\Delta_l^{-1} |u(k)| \sin pk \Big|_0^{\infty} \leq 0$$

Suppose $l\Delta_l^{-1} |u(k)| \sin pk \Big|_0^{2\pi} < 0$

$$ml\Delta_l^{-1} \sin pk \Big|_0^{\pi} + ml\Delta_l^{-1} \sin pk \Big|_{2\pi}^{4\pi} + \cdots$$

$$\rightarrow ml\Delta_l^{-1} |u(k)| \sin pk \Big|_0^{2\pi} < 0$$

The contradiction gives the

$$l\Delta_l^{-1} |u(k)| \sin pk \Big|_0^{\infty} = 0$$

Definition 12.2 *Let* $I_l^{(m)} = [ml, ml + a]$. *If* $l \sum_{m \in Z} \Delta_l^{-1} |u(k)|^2 \Big|_{ml}^{ml+a} < \infty$ *for some $a > 0$, then we say that $f \in L_l(-\infty, \infty)$ and we denoted* $l \sum_{m \in N(0)} \Delta_l^{-1} |u(k)|^2 \Big|_{ml}^{ml+a} = l\Delta_l^{-1} u(k)|_0^{\infty}$

Theorem 12.2 *If* $u(k) \in L_l(-\infty, \infty)$, *we have*

$$l\Delta_l^{-1} |u(k)| \left(\frac{1 - \cos pk}{k} \right) \Big|_{-\infty}^{\infty} = l\Delta_l^{-1} \left(\frac{|u(k)| - |u(-k)|}{k} \right) \Big|_0^{\infty} \tag{12.3}$$

Proof *Since* $\dfrac{1 - \cos pk}{k} = 0$ *at $k = 0$, it is bounded.*

Therefore, $l\Delta_l^{-1} |u(k)| \left(\dfrac{1 - \cos pk}{k} \right) \Big|_{-\infty}^{\infty}$ *exists as* $u \in L_l(-\infty, \infty)$.

By theorem (2.3) as u is bounded.

$$l\Delta_l^{-1} |u(k)| \cos pk|_0^{\infty} = 0 \tag{12.4}$$

Now we write,

$$l\Delta_l^{-1}|u(k)|\left(\frac{1-\cos pk}{k}\right)\Big|_{-\infty}^{\infty}$$

$$= l\Delta_l^{-1}|u(k)|\left(\frac{1-\cos pk}{k}\right)\Big|_0^{\infty} + l\Delta_l^{-1}|u(k)|\left(\frac{1-\cos pk}{k}\right)\Big|_{-\infty}^0$$

$$= l\Delta_l^{-1}\left[|u(k)|-|u(-k)|\right]\left(\frac{1-\cos pk}{k}\right)\Big|_{-\infty}^{\infty}$$

From Equation (12.4), we get the result.

Example 12.1 *Let $l = \dfrac{\pi}{N}, I = [0,2\pi], a < b$, and s_l is the orthonormal arrangement of geometrical capabilities depicted in Equation (2.3). The discrete Fourier series of $u(k)$ is characterized as*

$$u(k) = \frac{a_0}{2} + \sum_{v=1}^{V-1}(a_v \cos vk + b_v \sin vk) + \frac{a_N}{2}\cos Nk, \tag{12.5}$$

where the coefficients begin, given the following formulas:

$$a_0 = \frac{l}{\pi}\Delta_l^{-1}u(k)\Big|_0^{2\pi} = \frac{l}{\pi}\sum_{r=1}^{\left[\frac{2\pi}{l}\right]}u(2\pi - rl) \tag{12.6}$$

$$a_n = \frac{l}{\pi}\Delta_l^{-1}u(k)\cos nk\Big|_0^{2\pi} = \frac{l}{\pi}\sum_{r=1}^{\left[\frac{2\pi}{l}\right]}u(2\pi - rl)\cos(2\pi - rl) \tag{12.7}$$

$$b_n = \frac{l}{\pi}\Delta_l^{-1}u(k)\sin nk\Big|_0^{2\pi} = \frac{l}{\pi}\sum_{r=1}^{\left[\frac{2\pi}{l}\right]}u(2\pi - rl)\sin(2\pi - rl) \tag{12.8}$$

Consider the function $u(k) = k$ with Fourier coefficients

$$a_0 = \frac{l}{\pi}\Delta_l^{-1}k\Big|_0^{2\pi} = \frac{l}{\pi}\Delta_l^{-1}k^{(1)}\Big|_0^{2\pi} = \frac{l}{\pi}\left[\frac{k_l^{(2)}}{2l}\right]_0^{2\pi} \tag{12.9}$$

$$= \frac{l}{\pi}\left[\frac{(2\pi)_l^{(2)}}{2l} - \frac{(0)_l^{(2)}}{2l}\right] = \frac{l}{\pi}\left[\frac{2\pi(2\pi - l)}{2l}\right] = \frac{3\pi}{2}$$

$$a_n = \frac{l}{\pi}\Delta_l^{-1}u(k)\cos nk\Big|_0^{2\pi} = \frac{l}{\pi}\Delta_l^{-1}k\cos nk\Big|_0^{2\pi}$$

$$= \frac{l}{\pi}\left\{k\Delta_l^{-1}\cos nk - \Delta_l^{-1}\left[\Delta_l^{-1}\cos n(k+l)\Delta_l k\right]\right\}\Big|_0^{2\pi}$$

$$= \frac{l}{\pi}\left\{k\left[\frac{\cos n(k-l) - \cos nk}{2(1-\cos nl)}\right] - \Delta_l^{-1}\left[\frac{\cos nk - \cos n(k+l)}{2(1-\cos nl)}l\right]\right\}\Big|_0^{2\pi}$$

$$= \frac{l}{\pi}\left\{k\left[\frac{\cos n(k-l) - \cos nk}{2(1-\cos nl)}\right] - l\left[\frac{\cos n(k-l) - 2\cos nk + \cos n(k+l)}{[2(1-\cos nl)]^2}\right]\right\}\Big|_0^{2\pi}$$

$$
= \frac{l}{\pi} \left\{ (2\pi) \left[\frac{\cos n[(2\pi) - l] - \cos n[2\pi]}{2(1 - \cos nl)} \right] \right.
$$

$$
\left. - l \left[\frac{\cos n[(2\pi) - l] - 2\cos n[2\pi] + \cos n[(2\pi) + l]}{[2(1 - \cos nl)]^2} \right] \right\}
$$

$$
- \left\{ 0 - l \left[\frac{\cos n[0 - l] - 2\cos n[0] + \cos n[0 + l]}{[2(1 - \cos nl)]^2} \right] \right\}
$$

which gives,

$$
a_n = \frac{l}{\pi} \left\{ 2\pi \left(\frac{\cos n(2\pi - l) - \cos n2\pi}{2(1 - \cos nl)} \right) \right\} \tag{12.10}
$$

Similarly,

$$
b_n = \frac{l}{\pi} \Delta_l^{-1} u(k) \sin nk \Big|_0^{2\pi} = \frac{l}{\pi} \Delta_l^{-1} k \sin nk \Big|_0^{2\pi}
$$

$$
= \frac{l}{\pi} \left\{ k \Delta_l^{-1} \sin nk - \Delta_l^{-1} \left[\Delta_l^{-1} \sin n(k + l) \Delta_l k \right] \right\} \Big|_0^{2\pi}
$$

$$
= \frac{l}{\pi} \left\{ k \left[\frac{\sin n(k - l) - \sin nk}{2(1 - \cos nl)} \right] - l \left[\frac{\sin n(k - l) - 2\sin nk + \sin n(k + l)}{[2(1 - \cos nl)]^2} \right] \right\} \Big|_0^{2\pi}
$$

$$
= \frac{l}{\pi} \left\{ (2\pi) \left[\frac{\sin n[(2\pi) - l] - \sin n[2\pi]}{2(1 - \cos nl)} \right] \right.
$$

$$
\left. - l \left[\frac{\sin n[(2\pi) - l] - 2\sin n[2\pi] + \sin n[(2\pi) + l]}{[2(1 - \cos nl)]^2} \right] \right\}
$$

$$
- \left\{ 0 - l \left[\frac{\sin n[0 - l] - 2\sin n[0] + \sin n[0 + l]}{[2(1 - \cos nl)]^2} \right] \right\}
$$

which gives,

$$
b_n = \frac{l}{\pi} \left\{ 2\pi \left(\frac{\sin n(2\pi - l) - \sin n2\pi}{2(1 - \cos nl)} \right) \right\} \tag{12.11}
$$

Verification:　　*Consider (12.5) with* $l = \dfrac{\pi}{2}, k = \pi$ *and* $N = 2$,

From Equation (12.9), we have　　$a_0 = \dfrac{3\pi}{2}$

From Equation (12.10), we have

$$
a_1 = \frac{\pi}{2\pi} \left\{ (2\pi) \left(\frac{\cos\left(2\pi - \frac{\pi}{2}\right) - \cos(2\pi)}{2\left(1 - \cos\left(\frac{\pi}{2}\right)\right)} \right) \right\}
$$

$$
= \frac{1}{2} \left\{ (2\pi) \left(\frac{\cos\left(\frac{3\pi}{2}\right) - \cos(2\pi)}{2\left(1 - \cos\left(\frac{\pi}{2}\right)\right)} \right) \right\}
$$

$$
= \frac{1}{2} \left\{ (2\pi) \left(\frac{0 - 1}{2(1 - 0)} \right) \right\} = -\frac{\pi}{2}
$$

$$
a_2 = \frac{1}{2} \left\{ (2\pi) \left(\frac{\cos(3\pi) - \cos(4\pi)}{2(1 - \cos(\pi))} \right) \right\}
$$

$$= \frac{1}{2}\left\{ (2\pi) \left(\frac{-1-1}{2(1+1)} \right) \right\} = -\frac{\pi}{2}$$

From Equation (12.11), we have

$$b_1 = \frac{\pi}{2\pi}\left\{ (2\pi) \left(\frac{\sin\left(2\pi - \dfrac{\pi}{2}\right) - \sin(2\pi)}{2\left(1 - \cos\left(\dfrac{\pi}{2}\right)\right)} \right) \right\}$$

$$= \frac{1}{2}\left\{ (2\pi) \left(\frac{\sin\left(\dfrac{3\pi}{2}\right) - \sin(2\pi)}{2\left(1 - \cos\left(\dfrac{\pi}{2}\right)\right)} \right) \right\}$$

$$= \frac{1}{2}\left\{ (2\pi) \left(\frac{-1-0}{2(1-0)} \right) \right\} = -\frac{\pi}{2}$$

Now LHS of Equation $(12.5) = 3.14159$

$$\text{RHS of Equation } (12.5) = \frac{a_0}{2} + \sum_{v=1}^{V-1}(a_v \cos vk + b_v \sin vk) + \frac{a_N}{2}\cos Nk$$

$$= \frac{3\pi}{4} + a_1 \cos \pi + b_1 \sin \pi + \frac{a_2}{2}\cos 2\pi$$

$$= \frac{3\pi}{4} - \frac{\pi}{2}\cos \pi - \frac{\pi}{2}\sin \pi - \frac{\pi}{4}\cos 2\pi$$

$$= 2.35619 + 1.57079 - 0 - 0.78539$$

$$= 3.14159$$

Consider the following example of discrete Fourier series for k^2

Example 12.2 *Consider* (12.5) *where* $u(k) = k^2$,

$$(i.e)\quad k^2 = \frac{a_0}{2} + \sum_{v=1}^{V-1}(a_V \cos vk + b_v \sin vk) + \frac{a_N}{2}\cos Vk \tag{12.12}$$

$$\text{where } a_0 = \frac{l}{\pi}\Delta_l^{-1}k^2 \Big|_0^{2\pi} = \frac{l}{\pi}\Delta_l^{-1}\left[lk_l^{(1)} + lk_l^{(2)} \right]\Big|_0^{2\pi}$$

$$= \frac{l}{2\pi}k(k-l) + \frac{1}{3\pi}k(k-l)(k-2l)\Big|_0\, 2\pi$$

$$= \frac{3\pi^2}{4} + \pi^2 = \frac{7\pi^2}{4}$$

$$a_n = \frac{l}{\pi}\Delta_l^{-1}k^2 \cos nk \Big|_0^{2\pi}$$

$$= \frac{l}{\pi}\Delta_l^{-1}k^2 \cos nk \Big|_0^{2\pi} - \Delta_l^{-1}\left[\Delta_l^{-1}\cos nk - \cos n(k+l)\Delta_l k^2 \right]$$

$$= \frac{l}{\pi}\left\{ k^2\left(\frac{\cos n(k-l) - \cos nk}{2(1 - \cos nl)} \right) \right.$$

$$\left. - \Delta_l^{-1}\left[\frac{\cos nk - \cos n(k+l)}{2(1 - \cos nl)}\Delta_l\left(lk_l^{(1)+k_l^{(2)}} \right) \right] \right\}$$

$$\Delta_l^{-1}\left[\frac{\cos nk - \cos n(k+l)}{2(1-\cos nl)}\left(l^2 + 2lk_l^{(1)}\right)\right]$$

$$= \Delta_l^{-1}\left[l^2\left(\frac{\cos nk - \cos n(k+l)}{2(1-\cos nl)}\right) + 2l\left(k\frac{\cos nk - \cos n(k+l)}{2(1-\cos nl)}\right)\right]$$

$$= l^2\Delta_l^{-1}\left(\frac{\cos nk - \cos n(k+l)}{2(1-\cos nl)}\right) + 2l\Delta_l^{-1}\left(k\frac{\cos nk - \cos n(k+l)}{2(1-\cos nl)}\right)$$

$$l^2\Delta_l^{-1}\left(\frac{\cos nk - \cos n(k+l)}{2(1-\cos nl)}\right)$$

$$= l^2\left(\frac{\dfrac{\cos n(k-l) - \cos nk}{2(1-\cos nl)} - \dfrac{\cos nk - \cos n(k+l)}{2(1-\cos nl)}}{2(1-\cos nl)}\right)$$

$$= l^2\left(\frac{\cos n(k-l) - \cos nk - \cos nk - \cos n(k+l)}{[2(1-\cos nl)]^2}\right)$$

$$= l^2\left(\frac{\cos n(k-l) - 2\cos nk + \cos n(k+l)}{[2(1-\cos nl)]^2}\right)$$

$$(2l)\Delta_l^{-1}\left(k\frac{\cos nk - \cos n(k-l)}{2(1-\cos nl)}\right) = k\left(\frac{\cos n(k-l) - 2\cos nk + \cos n(k+l)}{2(1-\cos nl)}\right)$$

$$-\Delta_l^{-1}\left(\Delta_l^{-1}\frac{\cos nk - \cos n(k-l)}{2(1-\cos nl)}\Delta_l k\right)$$

$$\Delta_l^{-1}\left(\frac{\cos nk - \cos n(k+l)}{2(1-\cos nl)}\right) = \left(\frac{\dfrac{\cos n(k-l) - \cos nk}{2(1-\cos nl)} - \dfrac{\cos nk - \cos n(k+l)}{2(1-\cos nl)}}{2(1-\cos nl)}\right)$$

$$= \left(\frac{\cos n(k-l) - \cos nk - \cos nk - \cos n(k+l)}{[2(1-\cos nl)]^2}\right)$$

$$= \left(\frac{\cos n(k-l) - 2\cos nk + \cos n(k+l)}{[2(1-\cos nl)]^2}\right)$$

$$\Delta_l^{-1}\left(\frac{\cos n(k-l)k - 2\cos nk + \cos n(k+l)}{[2(1-\cos nl)]^2}\right)$$

$$= \left\{\frac{\left(\frac{\cos n(k-2l) - \cos n(k-l)}{2(1-\cos nl)}\right) - 2\left(\frac{\cos n(k-l) - \cos nk}{2(1-\cos nl)}\right) + \left(\frac{\cos nk - \cos n(k+l)}{2(1-\cos nl)}\right)}{[2(1-\cos nl)]^2}\right\}$$

$$= \left(\frac{\cos n(k-2l) - \cos n(k-1) - 2\cos n(k-l) + 2\cos nk + \cos nk - \cos n(k+l)}{[2(1-\cos nl)]^3}\right)$$

$$= k\left(\frac{\cos n(k-l) - 2\cos nk + \cos n(k+l)}{[2(1-\cos nl)]^2}\right)$$

$$-l\left(\frac{\cos n(k-l) - 3\cos nk + 3\cos n(k+l) - \cos n(k+2l)}{[2(1-\cos nl)]^3}\right)$$

Now we get the equation

$$a_n = \frac{l}{\pi}\left\{k^2\left(\frac{\cos n(k-l)-\cos nk}{2(1-\cos nl)}\right)-\left(\frac{\cos n(k-l)-2\cos nk+\cos n(k+l)}{[2(1-\cos nl)]^2}\right)\right.$$

$$+2l\left[k\left(\frac{\cos n(k-l)-2\cos nk+\cos n(k+l)}{[2(1-\cos nl)]^3}\right)\right.$$

$$\left.\left.\left[+2l^2\left(\frac{\cos n(k-l)-3\cos nk+3\cos n(k+l)-\cos n(k+2l)}{[2(1-\cos nl)]^3}\right)\right]\right\}_0^{2\pi}$$

$$=\frac{l}{\pi}\left\{k^2\left(\frac{\cos n(k-l)-\cos nk}{2(1-\cos nl)}\right)\right.$$

$$-(l^2+2lk)\left(\frac{\cos n(k-l)-\cos nk-2\cos nk+\cos n(k+l)}{[2(1-\cos nl)]^2}\right)$$

$$\left.+2l^2\left(\frac{\cos n(k-l)-\cos nk-3\cos n(k+l)-\cos n(k+2l)}{[2(1-\cos nl)]^3}\right)\right\}_0^{2\pi}$$

$$a_n=\frac{l}{\pi}\left\{(2\pi)^2\left(\frac{\cos n[(2\pi)-l]-\cos[2\pi]}{2(1-\cos nl)}\right)-(l^2+2l[2\pi])\right.$$

$$\times\left(\frac{\cos n[(2\pi)-l]-2\cos[2\pi]+\cos n[(2\pi)+l]}{[2(1-\cos nl)]^2}\right)$$

$$\left.-\left(\frac{\cos n[(2\pi)-l]-3\cos n[2\pi]+3\cos n[(2\pi)+l]-\cos n[(2\pi)+2l]}{[2(1-\cos nl)]^3}\right)\right\}$$

$$-\left\{(0)\left(\frac{\cos n[0-l]-\cos n[0]}{2(1-\cos nl)}\right)\right.$$

$$-(l^2+2l(\hat{l}2\pi))\left(\frac{\cos n[0-l]-2\cos n[0]+\cos n[0-l]}{[2(1-\cos nl)]^2}\right)$$

$$\left.-\left(\frac{\cos n[0-l]-3\cos n[0]+\cos n[0-l]-\cos n[0-2l]}{[2(1-\cos nl)]^3}\right)\right\}$$

$$a_n=\frac{l}{\pi}\left\{4\pi^2\left(\frac{\cos n(2\pi-l)-\cos n2\pi}{2(1-\cos nl)}\right)\right.$$

$$\left.-4l\left(\frac{\cos n(2\pi-l)-2\cos n2\pi+\cos n(2\pi+l)}{[2(1-\cos nl)]^2}\right)\right\} \tag{12.13}$$

Similarly,

$$a_n=\frac{l}{\pi}\left\{4\pi^2\left(\frac{\sin n(2\pi-l)-\sin n2\pi}{2(1-\cos nl)}\right)\right.$$

$$\left.-4l\left(\frac{\sin n(2\pi-l)-2\sin n2\pi+\sin n(2\pi+l)}{[2(1-\cos nl)]^2}\right)\right\} \tag{12.14}$$

Verification: *Now taking $l=\dfrac{\pi}{2}$ in Equations (12.13), we get*

$$a_1=\frac{l}{2\pi}\left\{4\pi^2\left(\frac{\cos(2\pi-\frac{\pi}{2})-\cos 2\pi}{2(1-\cos\frac{\pi}{2})}\right)\right.$$

$$-\frac{4\pi^2}{2}\left(\frac{\cos(2\pi-\frac{\pi}{2})-\cos 2\pi+\cos(2\pi+\frac{\pi}{2})}{[2(1-\cos\frac{\pi}{2})]^2}\right)\Big\}$$

$$=\frac{l}{2}\left\{4\pi^2\left(\frac{\cos(\frac{3\pi}{2})-\cos 2\pi}{2(1-\cos\frac{\pi}{2})}\right)-2\pi^2\left(\frac{\cos(\frac{3\pi}{2})-2\cos 2\pi+\cos(\frac{5\pi}{2})}{[2(1-\cos\frac{\pi}{2})]^2}\right)\right\}$$

$$a_1=-\frac{\pi^2}{2}$$

$$a_2=\frac{l}{2}\left\{4\pi^2\left(\frac{\cos 2(\frac{3\pi}{2})-\cos 4\pi}{2(1-\cos 2(\frac{\pi}{2}))}\right)-2\pi^2\left(\frac{\cos(2(\frac{3\pi}{2}))-2\cos 4\pi+\cos 2(\frac{5\pi}{2})}{[2(1-\cos 2(\frac{\pi}{2}))]^2}\right)\right\}$$

$$=\frac{l}{2}\left\{4\pi^2\left(\frac{\cos 3\pi-\cos 4\pi}{2(1-\cos\pi)}\right)-2\pi^2\left(\frac{\cos 3\pi-2\cos 4\pi+\cos 5\pi}{[2(1-\cos\pi]^2}\right)\right\}$$

$$a_2=-\frac{3\pi^2}{4}$$

Now taking $l=\dfrac{\pi}{2}$ in Equation (12.14), we get

$$b_1=\frac{l}{2}\left\{4\pi^2\left(\frac{\sin 2(\frac{3\pi}{2})-\sin 4\pi}{2(1-\cos 2(\frac{\pi}{2}))}\right)-2\pi^2\left(\frac{\sin 2(\frac{3\pi}{2})-2\sin 4\pi+\sin 2(\frac{5\pi}{2})}{[2(1-\cos 2(\frac{\pi}{2}))]^2}\right)\right\}$$

$$=\frac{l}{2}\left\{4\pi^2\left(\frac{\cos 3\pi-\cos 4\pi}{2(1-\cos\pi)}\right)-2\pi^2\left(\frac{\cos 3\pi-2\cos 4\pi+\cos 5\pi}{[2(1-\cos\pi]^2}\right)\right\}$$

$$b_1=0$$

Taking $k=\pi, N=2$ in Equation (12.12), we get

$$LHS\ of\ Equation\ (12.12)=\pi^2=9.8696$$

$$RHS\ of\ Equation\ (12.12)=\frac{7\pi^2}{2}-\frac{\pi^2}{2}\cos\pi+0-\frac{3\pi^2}{8}\cos 2\pi$$

$$=8.63590+4.93480-3.70110$$

$$=9.8696$$

Similarly, we can find discrete Fourier series of $u(k)=k^3$, where

$$a_n=\frac{l}{\pi}\left\{k^3\left(\frac{\cos n(k-l)-\cos nk}{2(1-\cos nl)}\right)\right.$$

$$-(l^3+3l^2+3lk^2)\left(\frac{\cos n(k-l)-2\cos nk+\cos n(k+l)}{[2(1-\cos nl)]^2}\right)$$

$$+(6l^3+6l^2k)\left(\frac{\cos n(k-l)-3\cos nk+3\cos n(k+l)-\cos n(k+2l)}{[2(1-\cos nl)]^3}\right)$$

$$\left.-6l^3\left(\frac{\cos n(k-l)-4\cos nk+6\cos n(k+l)-4\cos n(k+2l)+\cos n(k+3l)}{[2(1-\cos nl)]^4}\right)\right\}_0^{2\pi}$$

$$a_n=\frac{l}{\pi}\left\{8\pi^3\left(\frac{\cos n(2\pi-l)-\cos n2\pi}{2(1-\cos nl)}\right)\right.$$

$$-(6l^2\pi+12l\pi^2)\left(\frac{\cos n(2\pi-l)-2\cos n2\pi+\cos n(2\pi+l)}{[2(1-\cos nl)]^2}\right)$$

$$+12l^2\pi\left(\frac{\cos n(2\pi-l)-3\cos n2\pi+3\cos n(k+l)-\cos n(2\pi+2l)}{[2(1-\cos nl)]^3}\right)\right\}$$

and

$$b_n=\frac{l}{\pi}\left\{k^3\left(\frac{\sin n(k-l)-\sin nk}{2(1-\cos nl)}\right)\right.$$

$$-(l^3+3l^2k+3lk^2)\left(\frac{\sin n(k-l)-2\sin nk+\sin n(k+l)}{[2(1-\cos nl)]^2}\right)$$

$$+(6l^3+6l^2k)\left(\frac{\sin n(k-l)-3\sin nk+3\sin n(k+l)-\sin n(k+2l)}{[2(1-\cos nl)]^3}\right)$$

$$\left.-6l^3\left(\frac{\sin n(k-l)-4\sin nk+6\sin n(k+l)-4\sin n(k+2l)+\sin n(k+3l)}{[2(1-\cos nl)]^4}\right)\right\}_0^{2\pi}$$

$$b_n=\frac{l}{\pi}\left\{8\pi^3\left(\frac{\sin n(2\pi-l)-\sin n2\pi}{2(1-\cos nl)}\right)\right.$$

$$-(6l^2\pi+12l\pi^2)\left(\frac{\sin n(2\pi-l)-2\sin n2\pi+\sin n(2\pi+l)}{[2(1-\cos nl)]^2}\right)$$

$$\left.+12l^2\pi\left(\frac{\sin n(2\pi-l)-3\sin n2\pi+3\sin n(k+l)-\sin n(2\pi+2l)}{[2(1-\cos nl)]^3}\right)\right\}$$

Theorem 12.3 *If nl is not a multiple of 2π, then*

$$\Delta_l^{-1}k^p\cos nk=\sum_{q=1}^{p}S_q^p\left[\sum_{j=0}^{p}\frac{(-1)^j(q)_1^{(j)}l^jk_l^{(q-j)}}{(2(1-\cos nl))^{j+1}}\left(\sum_{r=0}^{j+1}(-1)^{r\;j+1}C_r\cos n(k-l+rl)\right)\right]$$

$$\Delta_l^{-1}k^p\sin nk=\sum_{q=1}^{p}S_q^p\left[\sum_{j=0}^{p}\frac{(-1)^j(q)_1^{(j)}l^jk_l^{(q-j)}}{(2(1-\cos nl))^{j+1}}\left(\sum_{r=0}^{j+1}(-1)^{r\;j+1}C_r\sin n(k-l+rl)\right)\right]$$

Proof *The proof follows by continuing the procedure of example (2.7).*

Theorem 12.4 $\Delta_l^{-1}k_l^{(m)}\cos nk=\displaystyle\sum_{j=0}^{m}\frac{(-1)^j(m)_1^{(j)}l^jk_l^{(m-j)}}{(2(1-\cos nl))^{j+1}}\left(\sum_{r=0}^{j+1}(-1)^{rj+1}C_r\cos n(k-l+rl)\right)$

$$\Delta_l^{-1}k_l^{(m)}\sin nk=\sum_{j=0}^{m}\frac{(-1)^j(m)_1^{(j)}l^jk_l^{(m-j)}}{(2(1-\cos nl))^{j+1}}\left(\sum_{r=0}^{j+1}(-1)^{rj+1}C_r\sin n(k-l+rl)\right)$$

Proof *The proof follows by Theorem (2.8) and* $k_l^{(m)}=\displaystyle\sum_{r=1}^{n}S_1^nl^{n-r}k^r$.

Theorem 12.5 *Discrete Fourier series for k^p*

$$k^p\approx\frac{a_0}{2}+\sum_{v=1}^{V-1}(g_v\cos vk+h_v\sin vk)+\frac{a_V}{2}\cos Vk,\text{ where }g_v\text{ and }h_v\text{ are given in Theorem (2.8)}$$

Theorem 12.6 *Discrete Fourier series for $k_l^{(p)}$*

$$k^p\approx\frac{a_0}{2}+\sum_{v=1^V-1}(a_v\cos vk+b_v\sin vk)+\frac{a_V}{2}\cos Vk,\text{ where }g_v\text{ and }h_v\text{ are given in Theorem (2.9)}$$

Similarly, one can find discrete Fourier series for several functions of $\dfrac{u(k)}{v(k)}$ like $\dfrac{a^k}{k_l^{(m)}}, \dfrac{e^k}{k}, \dfrac{a^k}{k^m}, \dfrac{e^k}{k_l^{(m)}}$ etc.

12.3   FOURIER STIRLING NUMBERS

In this chapter, we generate Fourier Stirling numbers that will be utilized to acquire the discrete Fourier series of the capability $u(k) = k_p$ with the notation F. The Stirling numbers of the first kind is used for expressing the polynomial factorial into polynomial, and Stirling number of second kind is used for expressing the polynomial into polynomial factorial. Here we generate Fourier Stirling numbers that gives the Fourier coefficients.

12.3.1   STIRLING NUMBER TABLE FOR FIRST KIND

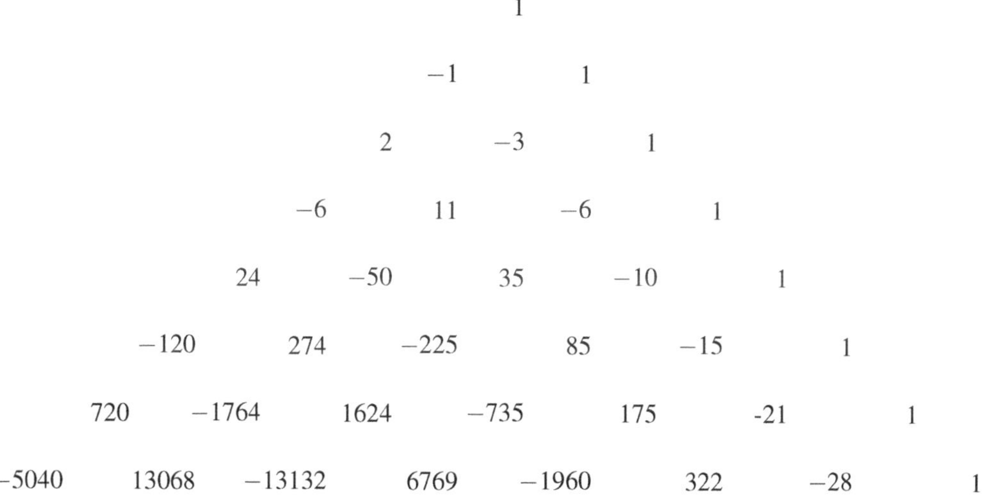

$$
\begin{array}{rrrrrrrrr}
 & & & & 1 & & & & \\
 & & & -1 & & 1 & & & \\
 & & 2 & & -3 & & 1 & & \\
 & -6 & & 11 & & -6 & & 1 & \\
24 & & -50 & & 35 & & -10 & & 1 \\
-120 & & 274 & & -225 & & 85 & & -15 & & 1 \\
720 & & -1764 & & 1624 & & -735 & & 175 & & -21 & & 1 \\
-5040 & & 13068 & & -13132 & & 6769 & & -1960 & & 322 & & -28 & & 1
\end{array}
$$

12.3.2 STIRLING NUMBER TABLE FOR SECOND KIND

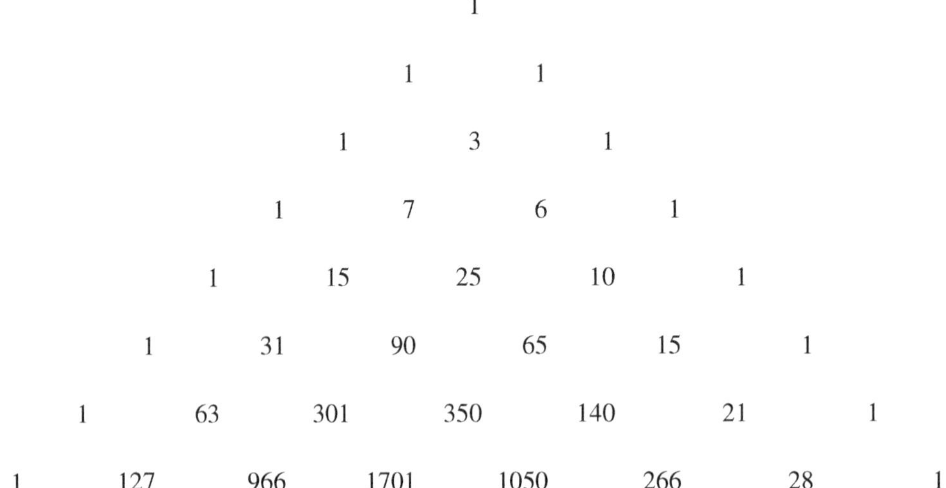

1							
1	1						
1	3	1					
1	7	6	1				
1	15	25	10	1			
1	31	90	65	15	1		
1	63	301	350	140	21	1	
1	127	966	1701	1050	266	28	1

12.3.3 FOURIER SERIES BY FOURIER STIRLING NUMBER

Consider the example (2.6). When $u(k) = k$, (12.6) becomes

$$a_0 = \frac{l}{\pi}\Delta_l^{-1}k = \frac{l}{\pi}\Delta_l^{-1}k_l^{(1)}\Big|_0^{2\pi} = \frac{l}{\pi}\left[\frac{k_l^{(1)}}{2l}\right]_0^{2\pi}$$

$$= \frac{l}{\pi}\left[\frac{(2\pi)_l^{(2)}}{2l} - \frac{(0)_l^{(2)}}{2l}\right]$$

$$= \frac{1}{2\pi}\left[(2\pi)_l^{(2)} - (0)_l^{(2)}\right] = \frac{3\pi}{2} \tag{12.15}$$

$$a_0 = \frac{l}{\pi}\Delta_l^{-1}u(k)\cos nk\Big|_0^{2\pi} = \frac{l}{\pi}\Delta_l^{-1}k\cos nk\Big|_0^{2\pi}$$

$$= \frac{l}{\pi}\left\{k\Delta_l^{-1}\cos nk\Delta_l^{-1}\left[\Delta_l^{-1}\cos n(k+l)\Delta_l F_1^1 k\right]\right\}\Big|_0^{2\pi}$$

$$= \frac{l}{\pi}\left\{k\left[\frac{\cos n(k-l)-\cos nk}{2(1-\cos nl)}\right] - \Delta_l^{-1}l\left[\frac{\cos nk-\cos n(k+l)}{2(1-\cos nl)}F_1^1 l\right]\right\}\Big|_0^{2\pi}$$

$$= \frac{l}{\pi}\left\{k\left[\frac{\cos n(k-l)-\cos nk}{2(1-\cos nl)}\right] - F_1^1 l\left[\frac{\cos n(k-l)-\cos nk-\cos nk+\cos n(k+l)}{[2(1-\cos nl)]^2}\right]\right\}$$

$$= \frac{l}{\pi}\left\{k\left[\frac{\cos n(k-l)-\cos nk}{2(1-\cos nl)}\right] - F_1^1 l\left[\frac{\cos n(k-l)-2\cos nk+\cos n(k+l)}{[2(1-\cos nl)]^2}\right]\right\}$$

$$a_n = \frac{l}{\pi}\left\{(2\pi)\left[\frac{\cos n[(2\pi)-l]-\cos n[2\pi]}{2(1-\cos nl)}\right]\right.$$

$$-F_1^1 l \left[\frac{\cos n[(2\pi) - l] - 2\cos n[2\pi] + \cos n[(2\pi) + l]}{[2(1 - \cos nl)]^2} \right] \Big\}$$

$$- \left\{ 0 \frac{\cos n(0 - l) - \cos n(0)}{2(1 - \cos nl)} - F_1^1 l \left[\frac{\cos n(0 - l) - 2\cos n[0] + \cos n[0 + l]}{[2(1 - \cos nl)]^2} \right] \right\}$$

$$a_n = \frac{l}{\pi} \left\{ 2\pi \left(\frac{\cos n(2\pi - l) - \cos n2\pi}{2(1 - \cos nl)} \right) \right\} \tag{12.16}$$

$$a_0 = \frac{l}{\pi} \Delta_l^{-1} u(k) \sin nk \Big|_0^{2\pi} = \frac{l}{\pi} \Delta_l^{-1} k \sin nk \Big|_0^{2\pi}$$

$$= \frac{l}{\pi} \left\{ k \Delta_l^{-1} \sin nk \Delta_l^{-1} \left[\Delta_l^{-1} \sin n(k + l) \Delta_l F_1^1 k \right] \right\} \Big|_0^{2\pi}$$

$$= \frac{l}{\pi} \left\{ k \left[\frac{\sin n(k - l) - \sin nk}{2(1 - \cos nl)} \right] - \Delta_l^{-1} l \left[\frac{\sin nk - \sin n(k + l)}{2(1 - \cos nl)} F_1^1 l \right] \right\} \Big|_0^{2\pi}$$

$$= \frac{l}{\pi} \left\{ k \left[\frac{\sin n(k - l) - \sin nk}{2(1 - \cos nl)} \right] - F_1^1 l \left[\frac{\sin n(k - l) - \sin nk - \sin nk + \sin n(k + l)}{[2(1 - \cos nl)]^2} \right] \right\}$$

$$= \frac{l}{\pi} \left\{ k \left[\frac{\sin n(k - l) - \sin nk}{2(1 - \cos nl)} \right] - F_1^1 l \left[\frac{\sin n(k - l) - 2\sin nk + \sin n(k + l)}{[2(1 - \cos nl)]^2} \right] \right\}$$

$$a_n = \frac{l}{\pi} \left\{ (2\pi) \left[\frac{\sin n[(2\pi) - l] - \sin n[2\pi]}{2(1 - \cos nl)} \right] \right.$$

$$\left. -F_1^1 l \left[\frac{\sin n[(2\pi) - l] - 2\sin n[2\pi] + \sin n[(2\pi) + l]}{[2(1 - \cos nl)]^2} \right] \right\}$$

$$- \left\{ 0 \frac{\sin n(0 - l) - \sin n(0)}{2(1 - \cos nl)} - F_1^1 l \left[\frac{\sin n(0 - l) - 2\sin n[0] + \sin n[0 + l]}{[2(1 - \cos nl)]^2} \right] \right\}$$

$$a_n = \frac{l}{\pi} \left\{ 2\pi \left(\frac{\sin n(2\pi - l) - \sin n2\pi}{2(1 - \cos nl)} \right) \right\} \tag{12.17}$$

Verification: Taking $l = \dfrac{\pi}{2}$ and $N = 2$ in Equations (12.5), we get

$$k = \frac{a_0}{2} + \sum_{v=1}^{V-1} (g_v \cos vk + h_v \sin nk) + \frac{g_N}{2} \cos Vk \tag{12.18}$$

where from Equation (12.9), $\quad a_0 = \dfrac{3\pi}{2}$

and from Equation (12.10),

$$a_1 = \frac{\pi}{2\pi} \left\{ (2\pi) \left(\frac{\cos \left(2\pi - \frac{\pi}{2}\right) - \cos(2\pi)}{2(1 - \cos(\frac{\pi}{2}))} \right) \right\}$$

$$= \frac{1}{2} \left\{ (2\pi) \left(\frac{\cos \left(\frac{3\pi}{2}\right) - \cos(2\pi)}{2(1 - \cos(\frac{\pi}{2}))} \right) \right\}$$

$$= \frac{1}{2}\left\{ (2\pi)\left(\frac{0-1}{2(1-0)} \right) \right\} = -\frac{\pi}{2}$$

$$a_2 = \frac{\pi}{2\pi}\left\{ (2\pi)\left(\frac{\cos(3\pi) - \cos(4\pi)}{2(1 - \cos(\pi))} \right) \right\}$$

$$= \frac{1}{2}\left\{ (2\pi)\left(\frac{-1-1}{2(1+1)} \right) \right\} = -\frac{\pi}{2}$$

From Equation (12.11), we have

$$b_1 = \frac{\pi}{2\pi}\left\{ (2\pi)\left(\frac{\sin\left(2\pi - \frac{\pi}{2}\right) - \sin(2\pi)}{2(1 - \cos(\frac{\pi}{2}))} \right) \right\}$$

$$= \frac{1}{2}\left\{ (2\pi)\left(\frac{\sin\left(\frac{3\pi}{2}\right) - \sin(2\pi)}{2(1 - \cos(\frac{\pi}{2}))} \right) \right\}$$

$$= \frac{1}{2}\left\{ (2\pi)\left(\frac{0-1}{2(1-0)} \right) \right\} = -\frac{\pi}{2}$$

Now taking $k = \pi$ and substituting $a_0 = \frac{3\pi}{2}, a_1 = -\frac{\pi}{2}, a_2 = -\frac{\pi}{2}$ and $b_1 = -\frac{\pi}{2}$ in (12.18)

LHS of Equation $(12.18) = \pi = 3.14159$

RHS of Equation $(12.18) = \dfrac{a_0}{2} \displaystyle\sum_{n=1}^{N-1} (a_n \cos nk + b_n \sin nk) + \dfrac{a_N}{2}\cos Nk$

$$= \frac{3\pi}{4} + a_1 \cos\pi + b_1 \sin\pi - \frac{a_2}{2}\cos 2\pi$$

$$= \frac{3\pi}{4} - \frac{\pi}{2} + \cos\pi + \sin\pi - \frac{\pi}{4}\cos 2\pi$$

$$= 2.35619 + 1.57079 - 0 - 0.78539$$

$$= 3.14159$$

Theorem 12.7 *Taking $u(k) = k^2$ in Equation (12.5), we get*

$$(i.e) \quad k^2 \approx \frac{a_0}{2} + \sum_{v=1}^{V-1} (g_v \cos vk + h_v \sin vk) + \frac{a_V}{2}\cos Nk \tag{12.19}$$

$$where \ a_0 = \frac{l}{\pi}\Delta_l^{-1}k^2 \bigg|_0^{2\pi} = \frac{l}{\pi}\Delta_l^{-1}\left[lk_l^{(1)} + lk_l^{(2)} \right]\bigg|_0^{2\pi}$$

$$= \frac{l}{2\pi}k(k-l) + \frac{1}{3\pi}k(k-l)(k-2l)\bigg|_0^{2\pi}$$

$$= \frac{3\pi^2}{4} + \pi^2 = \frac{7\pi^2}{4} \tag{12.20}$$

$$a_n = \frac{l}{\pi}\Delta_l^{-1}k^2\cos nk\Big|_0^{2\pi} = \frac{l}{\pi}\Delta_l^{-1}k^2\cos nk - \Delta_l^{-1}[\Delta_l^{-1}\cos n(k+l)\Delta_l k^2]\Big|_0^{2\pi}$$

$$= \frac{l}{\pi}\left\{ k^2\left(\frac{\cos n(k-l)-\cos nk}{2(1-\cos nl)}\right) - \Delta_l^{-1}\left[\frac{\cos nk-\cos n(k+l)}{2(1-\cos nl)}\Delta_l^{-1}\left(F_1^2 lk_l^{(1)}+F_2^2 lk_l^{(2)}\right)\right]\right\}$$

$$\Delta_l^{-1}\left[\frac{\cos nk-\cos n(k+l)}{2(1-\cos nl)}\Delta_l^{-1}\left(F_1^2 lk_l^{(1)}+F_2^2 lk_l^{(2)}\right)\right]$$

$$= \Delta_l^{-1}\left[F_1^2 l^2\left(\frac{\cos nk-\cos n(k+l)}{2(1-\cos nl)}\right)+F_2^2 2l\left(k\frac{\cos nk-\cos n(k+l)}{2(1-\cos nl)}\right)\right]$$

$$= F_1^2 l^2\Delta_l^{-1}\left(\frac{\cos nk-\cos n(k+l)}{2(1-\cos nl)}\right)+F_2^2 2l\Delta_l^{-1}\left(k\frac{\cos nk-\cos n(k+l)}{2(1-\cos nl)}\right)$$

$$F_1^2 l^2\Delta_l^{-1}\left(\frac{\cos nk-\cos n(k+l)}{2(1-\cos nl)}\right)$$

$$= F_1^2 l^2\left(\frac{\dfrac{\cos n(k-l)-\cos nk}{2(1-\cos nl)}-\dfrac{\cos nk-\cos n(k+l)}{2(1-\cos nl)}}{2(1-\cos nl)}\right)$$

$$= F_1^2 l^2\left(\frac{\cos n(k-l)-\cos nk-\cos nk-\cos n(k+l)}{[2(1-\cos nl)]^2}\right)$$

$$= F_1^2 l^2\left(\frac{\cos n(k-l)-2\cos nk-\cos n(k+l)}{[2(1-\cos nl)]^2}\right)$$

$$(2l)\Delta_l^{-1}\left(k\frac{\cos nk-\cos n(k-l)}{2(1-\cos nl)}\right)$$

$$= F_1^2 lk\left(\frac{\cos n(k-l)-2\cos nk+\cos n(k+l)}{[2(1-\cos nl)]^2}\right)$$

$$-F_1^2 F_2^2 2l^2\left(\frac{\cos n(k-l)-3\cos nk+3\cos n(k+l)-\cos n(k+2l)}{[2(1-\cos nl)]^3}\right)$$

$$\Delta_l^{-1}\left(\frac{\cos n(k-l)-2\cos nk-\cos n(k+l)}{[2(1-\cos nl)]^2}\right)$$

$$= \left\{\frac{\left(\dfrac{\cos n(k-2l)-\cos n(k-l)}{2(1-\cos nl)}\right)-2\left(\dfrac{\cos n(k-l)-\cos nk}{2(1-\cos nl)}\right)+\left(\dfrac{\cos nk-\cos n(k+l)}{2(1-\cos nl)}\right)}{[2(1-\cos nl)]^2}\right\}$$

$$= \left(\frac{\cos n(k-2l)-\cos n(k-l)-2\cos n(k-l)+2\cos nk+\cos nk-\cos n(k+l)}{[2(1-\cos nl)]^3}\right)$$

$$= k\left(\frac{\cos n(k-l) - 2\cos nk + \cos n(k+l)}{[2(1-\cos nl)]^2}\right)$$

$$-l\left(\frac{\cos n(k-l) - 3\cos nk + \cos n(k+l) - \cos n(k+2l)}{[2(1-\cos nl)]^3}\right)$$

Now we get the equation,

$$a_n = \frac{l}{\pi}\left\{k^2\left(\frac{\cos n(k-l) - \cos nk}{2(1-\cos nl)}\right) - F_1^2 1 l^2\left(\frac{\cos n(k-l) - 2\cos nk + \cos n(k+l)}{[2(1-\cos nl)]^2}\right)\right.$$

$$-F_2^2 2lk\left[\left(\frac{\cos n(k-l) - 2\cos nk + \cos n(k+l)}{[2(1-\cos nl)]^2}\right)\right.$$

$$\left.\left.+F_1^1 1 F_2^2 2 l^2\left(\frac{\cos n(k-l) - 3\cos nk + 3\cos n(k+l) - \cos n(k+2l)}{[2(1-\cos nl)]^3}\right)\right]\right\}_0^{2\pi}$$

$$a_n = \frac{l}{\pi}\left\{(2\pi)^2\left(\frac{\cos n[(2\pi) - l] - \cos[2\pi]}{2(1-\cos nl)}\right) - (l^2 + 2l[2\pi])\right.$$

$$\times\left(\frac{\cos n[(2\pi) - l] - 2\cos[2\pi] + \cos n[(2\pi) + l]}{[2(1-\cos nl)]^2}\right)$$

$$\left.-\left(\frac{\cos n[(2\pi) - l] - 3\cos n[2\pi] + 3\cos n[(2\pi) + l] - \cos n[(2\pi) + 2l]}{[2(1-\cos nl)]^3}\right)\right\}$$

$$-\left\{(0)\left(\frac{\cos n[0-l] - \cos n[0]}{2(1-\cos nl)}\right)\right.$$

$$-(l^2 + 2l(\hat{1}2\pi))\left(\frac{\cos n[0-l] - 2\cos n[0] + \cos n[0-l]}{[2(1-\cos nl)]^2}\right)$$

$$\left.-\left(\frac{\cos n[0-l] - 3\cos n[0] + \cos n[0-l] - \cos n[0-2l]}{[2(1-\cos nl)]^3}\right)\right\}$$

$$a_n = \frac{l}{\pi}\left\{4\pi^2\left(\frac{\cos n(2\pi - l) - \cos n2\pi}{2(1-\cos nl)}\right)\right. \tag{12.21}$$

$$\left.-4l\pi\left(\frac{\cos n(2\pi - l) - 2\cos n2\pi + \cos n(2\pi + l)}{[2(1-\cos nl)]^2}\right)\right\}$$

Similarly,

$$b_n = \frac{l}{\pi}\left\{k^2\left(\frac{\sin n(k-l) - \sin nk}{2(1-\cos nl)}\right)\right.$$

$$-(F_1^2 1 l^2 + F_2^2 2lk)\left(\frac{\sin n(k-l) - 2\sin nk + \sin n(k+l)}{[2(1-\cos nl)]^2}\right)$$

$$-F_1^1 1 F_2^2 2l^2 \left(\frac{\sin n(k-l) - 3\sin nk + 3\sin n(k+l) - \sin n(k+2l)}{[2(1-\cos nl)]^3} \right) \Big\}_0^{2\pi}$$

$$= \frac{l}{\pi} \left\{ (2\pi)^2 \left(\frac{\sin n[(2\pi)-l] - \sin[2\pi]}{2(1-\cos nl)} \right) - (l^2 + 2l[2\pi]) \right.$$

$$\times \left(\frac{\sin n[(2\pi)-l] - 2\sin[2\pi] + \sin n[(2\pi)+l]}{[2(1-\cos nl)]^2} \right)$$

$$- \left(\frac{\sin n[(2\pi)-l] - 3\sin n[2\pi] + 3\sin n[(2\pi)+l] - \sin n[(2\pi)+2l]}{[2(1-\cos nl)]^3} \right) \right\}$$

$$- \left\{ (0) \left(\frac{\sin n[0-l] - \sin n[0]}{2(1-\cos nl)} \right) \right.$$

$$- (l^2 + 2l(\hat{l}2\pi)) \left(\frac{\sin n[0-l] - 2\sin n[0] + \sin n[0-l]}{[2(1-\cos nl)]^2} \right)$$

$$- \left(\frac{\sin n[0-l] - 3\sin n[0] + \sin n[0-l] - \sin n[0-2l]}{[2(1-\cos nl)]^3} \right) \right\}$$

$$b_n = \frac{l}{\pi} \left\{ 4\pi^2 \left(\frac{\sin n(2\pi - l) - \sin n2\pi}{2(1-\cos nl)} \right) \right. \tag{12.22}$$

$$\left. -4l\pi \left(\frac{\sin n(2\pi - l) - 2\sin n2\pi + \sin n(2\pi + l)}{[2(1-\cos nl)]^2} \right) \right\}$$

Verification: *Now taking* $l = \dfrac{\pi}{2}$ *in Equations (12.21) and (12.22), we get*

$$a_1 = \frac{l}{2\pi} \left\{ 4\pi^2 \left(\frac{\cos(2\pi - \frac{\pi}{2}) - \cos 2\pi}{2(1-\cos\frac{\pi}{2})} \right) - \frac{4\pi^2}{2} \left(\frac{\cos(2\pi - \frac{\pi}{2}) - \cos 2\pi + \cos(2\pi + \frac{\pi}{2})}{[2(1-\cos\frac{\pi}{2})]^2} \right) \right\}$$

$$= \frac{l}{2} \left\{ 4\pi^2 \left(\frac{\cos(\frac{3\pi}{2}) - \cos 2\pi}{2(1-\cos\frac{\pi}{2})} \right) - 2\pi^2 \left(\frac{\cos(\frac{3\pi}{2}) - 2\cos 2\pi + \cos(\frac{5\pi}{2})}{[2(1-\cos\frac{\pi}{2})]^2} \right) \right\}$$

$$a_1 = -\frac{\pi^2}{2}$$

$$a_2 = \frac{l}{2} \left\{ 4\pi^2 \left(\frac{\cos 2(\frac{3\pi}{2}) - \cos 4\pi}{2(1-\cos 2(\frac{\pi}{2}))} \right) - 2\pi^2 \left(\frac{\cos(2(\frac{3\pi}{2})) - 2\cos 4\pi + \cos 2(\frac{5\pi}{2})}{[2(1-\cos 2(\frac{\pi}{2}))]^2} \right) \right\}$$

$$= \frac{l}{2} \left\{ 4\pi^2 \left(\frac{\cos 3\pi - \cos 4\pi}{2(1-\cos\pi)} \right) - 2\pi^2 \left(\frac{\cos 3\pi - 2\cos 4\pi + \cos 5\pi}{[2(1-\cos\pi)]^2} \right) \right\}$$

$$a_2 = -\frac{3\pi^2}{4}$$

Now taking $l = \dfrac{\pi}{2}$ in Equation (12.14), we get

$$b_1 = \frac{l}{2}\left\{4\pi^2\left(\frac{\sin 2(\frac{3\pi}{2}) - \sin 4\pi}{2(1 - \cos 2(\frac{\pi}{2}))}\right) - 2\pi^2\left(\frac{\sin 2(\frac{3\pi}{2}) - 2\sin 4\pi + \sin 2(\frac{5\pi}{2})}{[2(1 - \cos 2(\frac{\pi}{2}))]^2}\right)\right\}$$

$$= \frac{l}{2}\left\{4\pi^2\left(\frac{\cos 3\pi - \cos 4\pi}{2(1 - \cos \pi)}\right) - 2\pi^2\left(\frac{\cos 3\pi - 2\cos 4\pi + \cos 5\pi}{[2(1 - \cos \pi)]^2}\right)\right\}$$

$$b_1 = 0$$

Taking $k = \pi, N = 2$ in Equation (12.19), we get

$$(i.e) \quad k^2 = \frac{a_0}{2} + \sum_{v=1}^{V-1}(g_v \cos vk + h_v \sin vk) + \frac{a_N}{2}\cos Nk$$

$$LHS \ of \ Equation(12.12) = \pi^2 = 9.8696$$

$$RHS \ of \ Equation(12.12) = \frac{7\pi^2}{2} - \frac{\pi^2}{2}\cos \pi + 0 - \frac{3\pi^2}{8}\cos 2\pi$$

$$= 8.63590 + 4.93480 - 3.70110$$

$$= 9.8696$$

Similarly, we can find discrete Fourier series of $u(k) = k^3$. In general, we have the Fourier Stirling numbers in the following theorem.

Theorem 12.8 *If nl is not a multiple of 2π, then*

$$\Delta_l^{-1}k^p \cos nk = \sum_{q=1}^{p} F_q^p l^{p-q}\left[\sum_{j=0}^{p}\frac{(-1)^j (q)_1^{(j)} l^j k_l^{(q-j)}}{(2(1 - \cos nl))^{j+1}}\left(\sum_{r=0}^{j+1}(-1)^{r \ j+1}C_r \cos n(k - l + rl)\right)\right]$$

$$\Delta_l^{-1}k^p \sin nk$$

$$= \sum_{q=1}^{p} F_q^p l^{p-q}\left[\sum_{j=0}^{p}\frac{(-1)^j (q)_1^{(j)} l^j k_l^{(q-j)}}{(2(1 - \cos nl))^{j+1}}\left(\sum_{r=0}^{j+1}(-1)^{r \ j+1}C_r \sin n(k - l + rl)\right)\right]$$

12.4 GENERALIZED FOURIER SERIES FOR RATIONAL FUNCTIONS

In this chapter, first time we introduce the Fourier series for rational functions $\dfrac{u(k)}{v(k)}$ using summation form solutions of $\Delta_l^{-1}\dfrac{u(k)}{v(k)}\cos nk$ and $\Delta_l^{-1}\dfrac{u(k)}{v(k)}\sin nk$. Also we give an example of Fourier series for the rational function $\dfrac{1}{k_q}$.

12.4.1 CLOSED FACTORIAL FUNCTION

In this segment, we present a few essential definitions and a few outcomes which will be valuable for the ensuing conversation.

Definition 12.3 *Let $u(k)$ be the genuine esteemed capability characterized on $(0,\infty)$, $n \in \mathbb{N}(1)$ and $l \in (0,\infty)$. Then, the closed functional factorial and its reciprocal are, respectively, defined as*

$$(u(k))_l^{[n]} = u(k)u(k-l)u(k-2l)\ldots u(k-(n-1)l) \tag{12.23}$$

and when $u(k-rl) \neq 0$ for $r = 0,1,2,\ldots,n-1$, we have

$$\frac{1}{(u(k))_l^{[n]}} = \frac{1}{u(k)u(k-l)u(k-2l)\ldots u(k-(n-1)l)}. \tag{12.24}$$

In particular, when $u(k) = k$ (12.23) becomes the generalized polynomial factorial

$$k_l^{(n)} = k(k-l)(k-2l)\ldots(k-(n-1)l). \tag{12.25}$$

and (12.24) becomes the generalized reciprocal polynomial factorial

$$\frac{1}{k_l^{(n)}} = \frac{1}{k(k-l)(k-2l)\ldots(k-(n-1)l)}. \tag{12.26}$$

Lemma 12.1 *If $\lim\limits_{k\to\infty} \Delta_l^{-1}u(k) = 0$, then*

$$\Delta_l^{-1}u(k)\big|_k^\infty = (-1)\sum_{r=0}^\infty u(k+rl). \tag{12.27}$$

Theorem 12.9 *If $\lim\limits_{k\to\infty} \Delta_l^{-1}u(k) = 0$ for $r = 1,2,\ldots,m$ and $k \in [ml,\infty)$, then*

$$\Delta_l^{-1}u(k)\big|_k^\infty = (-1)^2\sum_{r=0}^\infty u(k-ml+rl). \tag{12.28}$$

Proof *The proof follows by taking Δ_l^{-1} on Equation (12.27) for $(m-1)$ times.*

12.4.2 DISCRETE FOURIER SERIES FOR RATIONAL FUNCTIONS

Here, we derive Δ_l^1 on certain closed reasonable capabilities to track down the aggregate on boundless series of summed up shut judicious factorial and polynomial factorial capabilities.

Theorem 12.10 *For $n \in \mathbb{N}(1)$ and $l \in (0,\infty)$. Then*

$$\Delta_l^{-1}\left[u(k)_l^{[n]}\Delta_{(n+1)l}u(k-nl)\right] = [u(k)]_l^{[n+1]}. \tag{12.29}$$

Proof *From the Definition (4.1), we find*

$$\Delta_l(u(k))_l^{[n]} = (u(k))_l^{[n-1]}\Delta_{nl}u(k-(n-1)l). \tag{12.30}$$

The evidence follows taking Δ_l^{-1} on both sides and replace n by $n+1$ in Equation (12.30).

Theorem 12.11 *For $l \in (0,\infty), k \in [0,\infty)$ we have*

$$\Delta_l^{-1}\left(k^{[m]}_{(n+1)l}\right)^{[n]}_l m(n+1)l(k-(n+r)l)^{(m+1)}_{(n+1)l} = \left(k^{[m]}_{(n+1)l}\right)^{[n+1]}_l.$$

Proof *The evidence follows taking $u(k) = \left(k^{[}_{(n+1)l}n]\right)^{[n]}_l$ in (12.30).*

Theorem 12.12 *For $n \in \mathbb{N}(1)$ and $l \in (0,\infty)$. Then,*

$$\Delta_l^{-1}\left[\frac{\Delta_{(n+1)l}u(k-(n-1)l)}{(u(k))^{[n]}_l}\right] = \frac{1}{(u(k-l))^{[n-1]}_l} \tag{12.31}$$

Proof *From the Definition (4.1), we find*

$$\Delta_l \frac{1}{(u(k))^{[n]}_l} = -\frac{\Delta_{nl}(k-(n-1)l)}{(u(k+l))^{[n+1]}_l} \tag{12.32}$$

The evidence follows by taking Δ_l^{-1} on both sides and replace k by $k - l$ then replace n by $n - 1$ in Equation (12.32).

Corollary 12.1 *Let $l \in (0,\infty)$ and $((k-l)^3)^{[3]}_l \neq 0$. Then*

$$\sum_{r=0}^{\infty} \frac{((k+rl)^3)^{[4]}_l - ((k+rl-3l)^3)^{[4]}_l}{((k+rl)^3)^{[4]}_l} = \frac{1}{((k-l)^3)^{[3]}_l} \tag{12.33}$$

Proof *The evidence follows taking $u(k) = k^3$ in Equation (12.31)*

Example 12.3 *In taking $k = 5, l = 2$ in Equation (12.33) we get*

$$\sum_{r=0}^{\infty} \frac{((5+2r)^3)^{[4]}_2 - ((5+2r-6)^3)^{[4]}_2}{((5+2r)^3)^{[4]}_2} = \frac{1}{(9)^3)^{[3]}_2}$$

Theorem 12.13 *Let $f(k) = \dfrac{u(k)}{u(k)}$ and $v(k) \neq 0$ on $(0, 2\pi)$ then*

$$u(k) = \frac{a_0}{2} + \sum_{n=1}^{N-1}(a_n \cos nk + b_n \sin nk) + \frac{a_N}{2}\cos Nk, \tag{12.34}$$

where,

$$a_0 = \frac{l}{\pi}\Delta_l^{-1}\frac{u(k)}{u(k)}\Big|_0^{2\pi} = \frac{l}{\pi}\sum_{r=1}^{\left[\frac{2\pi}{l}\right]}\frac{u(2\pi - rl)}{v(2\pi - rl)}$$

$$a_n = \frac{l}{\pi}\Delta_l^{-1}\frac{u(k)}{u(k)}\cos nk\Big|_0^{2\pi} = \frac{l}{\pi}\sum_{r=1}^{\left[\frac{2\pi}{l}\right]}\frac{u(2\pi - rl)}{v(2\pi - rl)}\cos n(2\pi - rl)$$

$$b_n = \frac{l}{\pi}\Delta_l^{-1}\frac{u(k)}{u(k)}\sin nk\Big|_0^{2\pi} = \frac{l}{\pi}\sum_{r=1}^{\left[\frac{2\pi}{l}\right]}\frac{u(2\pi - rl)}{v(2\pi - rl)}\sin n(2\pi - rl)$$

For the particular case $u(k) = 1$ and $v(k) = k^q$, we have

$$a_0 = \frac{l}{\pi} \sum_{r=1}^{\left[\frac{2\pi}{l}\right]} \frac{1}{(2\pi - rl)^q}$$

Verfication: *Let $l = 0.5, q = 1$ and $k = \pi$ in Equation* (12.34)

$$a_0 = \frac{0.5}{\pi} \sum_{r=0}^{12} \frac{1}{(2\pi - r(0.5))^1} = 1.34932$$

$$a_1 = \frac{0.5}{\pi} \sum_{r=0}^{12} \frac{\cos 1(2\pi - r(0.5))}{(2\pi - r(0.5))^1} = 0.57184 \qquad a_2 = \frac{0.5}{\pi} \sum_{r=0}^{12} \frac{\cos 2(2\pi - r(0.5))}{(2\pi - r(0.5))^1} = 0.35164$$

$$a_3 = \frac{0.5}{\pi} \sum_{r=0}^{12} \frac{\cos 3(2\pi - r(0.5))}{(2\pi - r(0.5))^1} = 0.21539 \qquad a_4 = \frac{0.5}{\pi} \sum_{r=0}^{12} \frac{\cos 4(2\pi - r(0.5))}{(2\pi - r(0.5))^1} = 0.11214$$

$$a_5 = \frac{0.5}{\pi} \sum_{r=0}^{12} \frac{\cos 5(2\pi - r(0.5))}{(2\pi - r(0.5))^1} = 0.02474 \qquad a_6 = \frac{0.5}{\pi} \sum_{r=0}^{12} \frac{\cos 6(2\pi - r(0.5))}{(2\pi - r(0.5))^1} = -0.05521$$

$$a_7 = \frac{0.5}{\pi} \sum_{r=0}^{12} \frac{\cos 7(2\pi - r(0.5))}{(2\pi - r(0.5))^1} = -0.13327 \qquad a_8 = \frac{0.5}{\pi} \sum_{r=0}^{12} \frac{\cos 8(2\pi - r(0.5))}{(2\pi - r(0.5))^1} = -0.21444$$

$$a_9 = \frac{0.5}{\pi} \sum_{r=0}^{12} \frac{\cos 9(2\pi - r(0.5))}{(2\pi - r(0.5))^1} = -0.30494 \qquad a_{10} = \frac{0.5}{\pi} \sum_{r=0}^{12} \frac{\cos 10(2\pi - r(0.5))}{(2\pi - r(0.5))^1} = 0.41544$$

Now

$$\text{LHS of Equation } (12.34) := \frac{1}{k} = \frac{1}{\pi} = 0.31831$$

$$\text{RHS of Equation } (12.34) := \frac{1.34932}{2} - 0.57184 + 0.35164 - 0.21539$$

$$+ 0.11214 - 0.02474 - 0.05521 + 0.13327$$

$$- 0.21444 + 0.30494 - 0.41544 + 0.57163$$

$$= 0.31781$$

Example 12.4 *Letting $u(k) = 1$, $v(k) = k^2$ in $f(k) = \dfrac{u(k)}{v(k)}$ and $v(k) \neq 0$ on $(0, 2\pi)$, then the discrete Fourier series becomes,*

$$\frac{1}{k^p} = \frac{a_0}{2} + \sum_{v=1}^{V-1} (g_v \cos vk + h_v \sin vk) + \frac{a_N}{2} \cos Vk, \tag{12.35}$$

where, $\qquad a_0 = \dfrac{l}{\pi} \sum_{r=1}^{\left[\frac{2\pi}{l}\right]} \dfrac{u(2\pi - rl)}{v(2\pi - rl)}$

$$a_n = \frac{l}{\pi} \sum_{r=1}^{\left[\frac{2\pi}{l}\right]} \frac{u(2\pi - rl)}{v(2\pi - rl)^2} \cos n(2\pi - rl)$$

$$b_n = \frac{l}{\pi} \sum_{r=1}^{\left[\frac{2\pi}{T}\right]} \frac{u(2\pi - rl)}{v(2\pi - rl)^2} \sin n(2\pi - rl)$$

Verification: *Letting $l = 0.5, q = 2$ and $k = \dfrac{\pi}{2}$ in Equation* (12.34)

$$a_0 = \frac{0.5}{\pi} \sum_{r=1}^{12} \frac{1}{(2\pi - r(0.5))^2} = 2.49782 \qquad a_1 = \frac{0.5}{\pi} \sum_{r=1}^{12} \frac{\cos 1(2\pi - r(0.5))}{(2\pi - r(0.5))^2} = 2.05171$$

$$a_2 = \frac{0.5}{\pi} \sum_{r=1}^{12} \frac{\cos 2(2\pi - r(0.5))}{(2\pi - r(0.5))^2} = 1.56917 \qquad a_3 = \frac{0.5}{\pi} \sum_{r=0}^{12} \frac{\cos 3(2\pi - r(0.5))}{(2\pi - r(0.5))^2} = 1.09652$$

$$a_4 = \frac{0.5}{\pi} \sum_{r=1}^{12} \frac{\cos 4(2\pi - r(0.5))}{(2\pi - r(0.5))^2} = 0.63529 \qquad a_5 = \frac{0.5}{\pi} \sum_{r=0}^{12} \frac{\cos 5(2\pi - r(0.5))}{(2\pi - r(0.5))^2} = 0.18654$$

$$a_6 = \frac{0.5}{\pi} \sum_{r=1}^{12} \frac{\cos 6(2\pi - r(0.5))}{(2\pi - r(0.5))^2} = 0.24833 \qquad a_7 = \frac{0.5}{\pi} \sum_{r=0}^{12} \frac{\cos 7(2\pi - r(0.5))}{(2\pi - r(0.5))^1} = -0.66738$$

$$a_8 = \frac{0.5}{\pi} \sum_{r=1}^{12} \frac{\cos 8(2\pi - r(0.5))}{(2\pi - r(0.5))^2} = -1.06771 \qquad a_9 = \frac{0.5}{\pi} \sum_{r=0}^{12} \frac{\cos 9(2\pi - r(0.5))}{(2\pi - r(0.5))^1} = -1.44489$$

$$a_10 = \frac{0.5}{\pi} \sum_{r=1}^{12} \frac{\cos 10(2\pi - r(0.5))}{(2\pi - r(0.5))^2} = -1.79159 \qquad a_{11} = \frac{0.5}{\pi} \sum_{r=0}^{12} \frac{\cos 11(2\pi - r(0.5))}{(2\pi - r(0.5))^1} = -2.09404$$

$$a_12 = \frac{0.5}{\pi} \sum_{r=1}^{12} \frac{\cos 10(2\pi - r(0.5))}{(2\pi - r(0.5))^2} = -2.31876$$

similarly,

$$b_1 = \frac{0.5}{\pi} \sum_{r=1}^{12} \frac{1}{(2\pi - r(0.5))^2} = 0.87886 \qquad b_2 = \frac{0.5}{\pi} \sum_{r=1}^{12} \frac{\cos 1(2\pi - r(0.5))}{(2\pi - r(0.5))^2} = 1.32376$$

$$b_3 = \frac{0.5}{\pi} \sum_{r=1}^{12} \frac{\cos 2(2\pi - r(0.5))}{(2\pi - r(0.5))^2} = 1.59359 \qquad b_4 = \frac{0.5}{\pi} \sum_{r=0}^{12} \frac{\cos 3(2\pi - r(0.5))}{(2\pi - r(0.5))^2} = 1.74458$$

$$b_5 = \frac{0.5}{\pi} \sum_{r=1}^{12} \frac{\cos 4(2\pi - r(0.5))}{(2\pi - r(0.5))^2} = 1.80068 \qquad b_6 = \frac{0.5}{\pi} \sum_{r=0}^{12} \frac{\cos 5(2\pi - r(0.5))}{(2\pi - r(0.5))^2} = 1.77346$$

$$b_7 = \frac{0.5}{\pi} \sum_{r=1}^{12} \frac{\cos 6(2\pi - r(0.5))}{(2\pi - r(0.5))^2} = 1.66762 \qquad b_8 = \frac{0.5}{\pi} \sum_{r=0}^{12} \frac{\cos 7(2\pi - r(0.5))}{(2\pi - r(0.5))^1} = 1.48268$$

$$b_9 = \frac{0.5}{\pi} \sum_{r=1}^{12} \frac{\cos 8(2\pi - r(0.5))}{(2\pi - r(0.5))^2} = 1.21284 \qquad b_{10} = \frac{0.5}{\pi} \sum_{r=0}^{12} \frac{\cos 9(2\pi - r(0.5))}{(2\pi - r(0.5))^1} = 0.84442$$

$$b_{11} = \frac{0.5}{\pi} \sum_{r=1}^{12} \frac{\cos 10(2\pi - r(0.5))}{(2\pi - r(0.5))^2} = 0.34814 \qquad b_{12} = \frac{0.5}{\pi} \sum_{r=0}^{12} \frac{\cos 11(2\pi - r(0.5))}{(2\pi - r(0.5))^1} = -0.35423$$

Therefore we have,

$$LHS\ of\ Equation\ (12.35) := \frac{1}{k^2} = \frac{1}{(\frac{\pi}{2})^2} = 0.40528$$

$$RHS\ of\ Equation\ (12.35) := \frac{2.49782}{2} - 0.87886 - 1.56917 - 1.59359$$

$$+0.635291 + 1.80068 + 0.24833$$

$$-1.66762 - 1.06771 + 1.21284 + 1.79153$$

$$= 0.40942$$

Theorem 12.14 *Let* $f(k) = \dfrac{u(k)^p}{u(k)^q}$ *and* $v(k)^q \neq 0$ *on* $(0, 2\pi)$ *then* $\quad \dfrac{1}{k^p} = \dfrac{a_0}{2} + \sum\limits_{n=1}^{N-1}(a_n \cos nk +$

$b_n \sin nk) + \dfrac{a_N}{2} \cos Nk,$

where, $\qquad a_0 = \dfrac{l}{\pi} \Delta_l^{-1} \dfrac{u(k)^p}{u(k)^q}\Big|_0^{2\pi} = \dfrac{l}{\pi} \sum\limits_{r=1}^{\left[\frac{2\pi}{l}\right]} \dfrac{u(2\pi - rl)^p}{v(2\pi - rl)^q}$

$$a_n = \dfrac{l}{\pi} \Delta_l^{-1} \dfrac{u(k)^p}{u(k)^q} \cos nk\Big|_0^{2\pi} = \dfrac{l}{\pi} \sum\limits_{r=1}^{\left[\frac{2\pi}{l}\right]} \dfrac{u(2\pi - rl)^p}{v(2\pi - rl)^q} \cos n(2\pi - rl)$$

$$b_n = \dfrac{l}{\pi} \Delta_l^{-1} \dfrac{u(k)^p}{u(k)^q} \sin nk\Big|_0^{2\pi} = \dfrac{l}{\pi} \sum\limits_{r=1}^{\left[\frac{2\pi}{l}\right]} \dfrac{u(2\pi - rl)^p}{v(2\pi - rl)^q} \sin n(2\pi - rl)$$

12.5 CONCLUSION

When $l \to 0$, discrete Fourier transform becomes usual Fourier series and the Fourier transforms. If $f(\cdot)dx$ does not exist, then we can replace $f(\cdot)$ by $l\Delta_l^{-1}(\cdot)$, and we can get several applications using discrete Fourier transform and its series using summation form of Δ_l^{-1}. Fourier series cannot be found for rational functions. Here we find discrete Fourier series for rational functions. As $l \to 0$, we can obtain Fourier series from discrete Fourier series.

BIBLIOGRAPHY

1. Ablowitz, M.J., & Ladik, J.F., On the solution fo a class of nonlinear partial difference equations. *Studies in Applied Mathematics*, 57 (1977), 1–12.
2. Britto Antony Xavier, G., Gerly, T.G., & Nasira Begum, H., Fine series of polynominals and polynomial factorials arising from generalized q-difference operator. *Far East Journal of Mathematical Sciences*, 94(1) (2014), 47–63.
3. Britto Antony Xavier, G., John Borg, S., & Meganathan, M., Discrete heat equation model with shift values. *Applied Mathematics*, 8 (2017), 1343–1350.

4. Bastos, N.R.O., Ferreira, R.A.C., & Torres, D.F.M., Discrete-time fractional variational problems. *Signal Processing*, 91(3) (2011), 513–524.

5. Proprnda, J., & Szmanda, B., On the oscillation of solutions of certain difference equation. *Demonstration Mathematica*, XVII(1) (1984), 153–164.

6. Maria Susai Manuel, M., Chandrasekar, V., & Britto Antony Xavier, G., Solutions and applications of certain class of difference equation. *International Journal of Applied Mathematics*, 24(6) (2011), 943–945.

7. Maria Susai Manuel, M., Chandrasekar, V., Britto Antony Xavier, G., & Pugalarau, R., Theory and application of the generalized difference operator of the 'n' kind (part 1). *Demonstration Mathematics*, 45(1) (2012), 95–106.

8. Miller K.S and Ross B; "The Discrete Simple of the Riemann Librevike Partial Derivative". Horwood, Chichester, UK 1989.

Appendix

A.1 PERFORMANCE ANALYSIS OF REDUNDANT MACHINING SYSTEM UNDER GENERALIZED TRIADIC POLICY

In this appendix, we present the mathematical formulations and equations used in the performance analysis of the redundant machining system with multiple working vacations under the generalized triadic policy, as discussed in Chapter 1.

ARRIVAL PROCESS:

The arrival process in the redundant machining system with multiple working vacations can be modeled using a Poisson process. Let λ denote the arrival rate of jobs.

SERVICE TIME DISTRIBUTION:

The service time distribution represents the time taken to process a job on a machine. It is often assumed to follow an exponential distribution with mean service time $1/\mu$, where μ is the mean service rate.

PERFORMANCE MEASURES:

Several performance measures are used to evaluate the efficiency and effectiveness of the redundant machining system. These include the following:

Average number of jobs in the system (Ls)
Average number of jobs in the queue (Lq)
Average waiting time in the system (Ws)
Average waiting time in the queue (Wq)
System throughput
System utilization

OPTIMIZATION TECHNIQUES:

Optimization techniques, such as mathematical programming, heuristic algorithms, and simulation-based optimization, can be applied to find optimal values for system parameters (e.g., threshold levels, vacation durations) that maximize system performance measures or minimize system costs (e.g., waiting times and idle times).

By incorporating these mathematical formulations and analysis methodologies, researchers can conduct a thorough investigation into the performance of redundant machining systems with multiple working vacations under the generalized triadic policy, leading to insights for system design, operation, and optimization strategies.

A.2 SOLVING PHYSICAL SYSTEM PROBLEMS USING PARTIAL DIFFERENTIAL EQUATIONS

EQUATIONS OF INTEREST:

This section introduces several fundamental partial differential equations that regulate physical systems:

Poisson's equation: A key equation in electrostatics and gravitation, Poisson's equation describes the distribution of mass or charge in a given region of space, leading to gravitational or electric fields, respectively.

Laplace's equation: Laplace's equation is a special case of Poisson's equation where the source term is zero. It describes systems in equilibrium, such as steady-state heat conduction or electrostatics without charge.

Wave equation: The wave equation governs the propagation of waves, including mechanical waves (e.g., sound waves), electromagnetic waves, and quantum mechanical waves.

Schrodinger's wave equation: A fundamental equation in quantum mechanics, Schrödinger's wave equation describes how the quantum state of a physical system changes over time.

Equation of heat flow: Also known as the heat equation, this PDE models the distribution of heat in a given region over time, describing phenomena such as heat conduction and diffusion.

SOLUTIONS OF PARTIAL DIFFERENTIAL EQUATIONS FOR SOME PHYSICAL SYSTEMS:

This subsection delves into specific physical systems and demonstrates how partial differential equations can be used to model and solve them:

Vibrating string: The behavior of a vibrating string can be described using the wave equation, leading to solutions that depict the string's displacement over time.

Propagation of sound in a gaseous medium: Sound waves traveling through a gaseous medium obey the wave equation, with additional considerations for properties such as density, pressure, and temperature variations.

A.3 SOLUTION TO ONE-DIMENSIONAL HEAT EQUATION WITH NON-LOCAL BOUNDARY CONDITIONS

This appendix provides supplementary information and analysis related to the chapter on the solution to the one-dimensional heat equation with non-local boundary conditions using a second-order accurate finite difference scheme.

STABILITY ANALYSIS:

Stability analysis is a crucial aspect of numerical methods to ensure that the solution does not exhibit unbounded growth or oscillations over time. More detailed explanations of stability analysis techniques, such as Von Neumann stability analysis for finite difference schemes, could be provided.

FINITE DIFFERENCE METHOD (FDM):

Finite difference methods are numerical techniques used to approximate solutions to differential equations by discretizing the domain into a grid of points. This section could include more detailed explanations of FDM, including forward difference, backward difference, and central difference schemes.

APPLICATION EXAMPLES:

Additional application examples demonstrating the solution of one-dimensional heat equations with non-local boundary conditions using finite difference methods could be provided. These examples could cover a range of physical scenarios to illustrate the versatility and effectiveness of the numerical approach.

FURTHER READING:

A list of references to relevant textbooks, research papers, and online resources for readers interested in exploring the topic in more detail could be included. This could encompass advanced topics in numerical methods, heat transfer, and computational physics.

A.4 IDENTIFICATION OF OPTIMUM PUBLIC ELECTRIC VEHICLE CHARGING STATIONS

CONSIDERATIONS FOR OPTIMAL CHARGING STATION PLACEMENT:

Additional considerations for optimal charging station placement may include factors such as:

Population density and traffic volume in surrounding areas.
Proximity to major highways, commercial centers, and residential areas.
Availability of amenities and services for electric vehicle (EV) users (e.g., rest areas and dining options).
Integration with existing infrastructure and urban planning initiatives.
Environmental impact and sustainability considerations.

INTEGRATION OF RENEWABLE ENERGY SOURCES:

Exploring the integration of renewable energy sources, such as solar panels or wind turbines, into EV charging stations could enhance sustainability and resilience. This could involve discussions on the feasibility, benefits, and challenges associated with renewable energy integration in charging infrastructure.

FUTURE RESEARCH DIRECTIONS:

This section could outline potential avenues for future research and development in the field of optimum public EV charging station identification. It may include topics such as advanced optimization algorithms, dynamic charging station management systems, and policy implications for promoting electric vehicle adoption and infrastructure deployment.

CASE STUDIES AND REAL-WORLD APPLICATIONS:

Including case studies or examples of real-world applications showcasing the implementation and impact of optimized public EV charging station networks could provide practical insights and illustrate the effectiveness of the proposed methodology.

A.5 SPANNING SOFT TREES

FURTHER DETAILS ON SPANNING SOFT TREES:

In this appendix, we provide additional information and insights into the concept of spanning soft trees, expanding upon the topics discussed in Chapter 5.

INTRODUCTION TO SPANNING SOFT TREES:

In this chapter, we introduced the notion of spanning soft trees as subgraphs of soft graphs that connect all nodes in the graph, forming a tree structure. These trees play a crucial role in various applications where uncertainty is inherent in the relationships between nodes.

PRELIMINARIES OF SOFT GRAPHS:

The preliminaries of soft graphs were discussed to provide a foundational understanding. Soft graphs extend traditional graph theory by incorporating uncertainty or fuzziness into edge weights or probabilities, allowing for more realistic modeling of real-world systems.

SPANNING SOFT TREES:

This chapter delved into the details of spanning soft trees, exploring algorithms and methodologies for finding these trees in soft graphs. We discussed optimization criteria for minimizing uncertainty or fuzziness associated with the tree edges, as well as practical applications in network design and decision-making.

FURTHER READING AND RESOURCES:

For readers interested in delving deeper into the topic of spanning soft trees and soft graph theory, we recommend exploring additional resources and research papers in the field. This could include works by leading researchers in soft graph theory and related areas of uncertainty modeling and optimization.

A.6   A REVIEW ON THE TRANSSHIPMENT PROBLEMS IN OPTIMIZATION THEORY

ADDITIONAL RESOURCES AND REFERENCES:

This appendix provides a list of additional resources and references for readers interested in further exploration of transshipment problems in optimization theory, building upon the discussions presented in Chapter 6.

Books:

"Network Flows: Theory, Algorithms, and Applications" by Ravindra K. Ahuja, Thomas L. Magnant, and James B. Orlin.
"Integer Programming" by Laurence A. Wolsey.
"Supply Chain Management: Strategy, Planning, and Operation" by Sunil Chopra and Peter Meindl.
"Handbook of Transportation Science" edited by Randolph Hall.

Online Resources:

Operations Research Society of America (ORSA) and The Institute for Operations Research and the Management Sciences (INFORMS): Websites offering access to journals, conferences, and educational materials related to operations research and optimization theory.
Scholarly Databases: Platforms such as Google Scholar, JSTOR, and IEEE Xplore provide access to a wide range of academic articles and research papers on transshipment problems and related topics.

Software and Tools:

Optimization software: Packages such as CPLEX, Gurobi, and AMPL offer optimization solvers that can be used to tackle transshipment problems and other optimization challenges.
Simulation Tools: Software like Arena and AnyLogic can be utilized to simulate and analyze complex supply chain and transportation systems, including transshipment operations.

Professional Organizations:

The Institute for Operations Research and the Management Sciences (INFORMS): An international professional society dedicated to advancing research and practice in operations research and analytics.

The Mathematical Optimization Society (MOS): A global organization promoting research and collaboration in mathematical optimization. These additional resources provide a comprehensive overview of transshipment problems and offer valuable insights for researchers, practitioners, and students interested in exploring this area further.

A.7 THE ANALYSIS OF SUSTAINABILITY OF WOMEN ENTREPRENEURSHIP BY THE USE OF COMPLEX FUZZY MATRICES

DETAILS OF COMPLEX FUZZY MATRICES CONSTRUCTION:

This appendix provides a detailed explanation of the construction and implementation of complex fuzzy matrices utilized in the analysis of the sustainability of women's entrepreneurship, as described in Chapter 8 of this study.

IDENTIFICATION OF CRITERIA:

The first step in constructing complex fuzzy matrices was the identification of criteria relevant to the sustainability of women's entrepreneurship.

Criteria such as economic performance, environmental impact, social responsibility, and innovation were selected based on their significance in assessing sustainability.

LINGUISTIC VARIABLES:

Each criterion was associated with linguistic variables representing different levels of achievement or performance. For example, linguistic terms such as "low," "medium," and "high" were defined to quantify the economic performance of women entrepreneurs.

EXPERT CONSULTATION:

Expert opinions were sought to determine the importance of each criterion and to assign weights accordingly.

Subject matter experts in the fields of entrepreneurship, sustainability, and fuzzy logic provided valuable insights to ensure the relevance and validity of the criteria.

DATA COLLECTION AND AGGREGATION:

Data about each criterion were collected from various sources, including surveys, interviews, and secondary research.

Quantitative data were aggregated and normalized to ensure consistency and comparability across different criteria. Quantitative data were aggregated and normalized to ensure consistency and comparability across different criteria.

CONSTRUCTION OF FUZZY MATRICES:

Fuzzy matrices were constructed to represent the relationships between criteria and linguistic variables.

Each row of the matrix corresponded to a criterion, while each column represented a linguistic variable.

Fuzzy membership functions were applied to quantify the degree of membership of each linguistic variable within its corresponding criterion.

A.8   FUZZY RANDOM MULTIOBJECTIVE QUADRATIC TRANSPORTATION PROBLEM

INFORMATION:

In this appendix, we provide supplementary details and resources related to the topic of the Fuzzy Random Multiobjective Quadratic Transportation Problem (FRMOQTP). This appendix serves to enhance the understanding of readers interested in exploring this specialized area further.

PROBLEM DEFINITION:

The FRMOQTP deals with the transportation of goods from multiple sources to multiple destinations, where various parameters such as transportation costs, demands, and supplies are subject to uncertainty.

Uncertainty in the problem is represented using both fuzzy and random parameters, making it a hybrid optimization problem that requires specialized techniques for solution.

SOLUTION APPROACHES:

Solving the FRMOQTP requires advanced optimization methods that can handle both fuzzy and random parameters simultaneously.

Metaheuristic algorithms such as genetic algorithms, particle swarm optimization, and simulated annealing have been adapted to handle the multiobjective nature and uncertainty in the problem. Fuzzy random programming techniques offer a systematic framework for modeling and solving optimization problems with fuzzy and random parameters, providing robust solutions under uncertainty.

APPLICATIONS AND CASE STUDIES:

The FRMOQTP has applications in diverse fields such as supply chain management, logistics, and transportation planning, where decision-making must account for uncertainty and multiple conflicting objectives.

Real-world case studies and application examples demonstrate the practical relevance of the FRMOQTP and the effectiveness of different solution approaches in addressing complex decision-making scenarios.

SOFTWARE AND TOOLS:

Various software tools and libraries such as MATLAB®, Python's SciPy, and AMPL can be utilized for modeling and solving the FRMOQTP.

Specialized packages for fuzzy logic and random optimization, such as MATLAB's Fuzzy Logic Toolbox and R's FuzzyToolkitUoN, offer dedicated functionalities for handling fuzzy and random parameters in optimization problems.

A.9 OPERATIONAL CONTROLLABILITY OF NEUTRAL HIGHER-ORDER INTEGRODIFFERENTIAL SYSTEMS

NUMERICAL METHODS AND SIMULATION TECHNIQUES:

Discussion on numerical methods and simulation techniques used for analyzing operational controllability could offer readers a deeper understanding of the computational aspects involved in the study. This section may cover simulation algorithms, software tools, and computational experiments conducted to validate theoretical results.

STABILITY ANALYSIS:

Stability analysis is often intertwined with controllability analysis in dynamical systems. This section could explore the stability properties of neutral higher-order integrodifferential systems and their relationship with operational controllability, providing insights into system behavior under various control strategies.

PRACTICAL CONSIDERATIONS AND APPLICATIONS:

Addressing practical considerations and real-world applications of operational controllability analysis could enhance the relevance of the study. This section may discuss engineering applications, control design implications, and potential benefits for system optimization and performance enhancement.

FUTURE RESEARCH DIRECTIONS:

Outlining future research directions and emerging topics in the field of operational controllability for neutral higher-order integrodifferential systems could inspire further investigation and innovation. This section may suggest areas for future exploration, methodological advancements, and interdisciplinary collaborations.

A.10 UNLEASHING THE POWER OF NATURE-INSPIRED OPTIMIZATION

NATURE-INSPIRED OPTIMIZATION ALGORITHMS:

This appendix provides an overview of the nature-inspired optimization algorithms discussed in Chapter 10.

GENETIC ALGORITHMS (GAS):

Genetic algorithms are inspired by the process of natural selection and genetics.
They involve a population of potential solutions (individuals) that evolve over successive generations through processes such as selection, crossover, and mutation.
GAs are widely used for solving optimization and search problems in various domains due to their ability to efficiently explore large solution spaces.

PARTICLE SWARM OPTIMIZATION (PSO):

Particle swarm optimization is inspired by the social behavior of bird flocks and fish schools.
In PSO, a population of particles moves through the search space, guided by their own best-known position and the global best-known position found by the swarm.
PSO algorithms are particularly effective for continuous optimization problems and have been successfully applied in engineering, economics, and other fields.

ANT COLONY OPTIMIZATION (ACO):

Ant colony optimization is inspired by the foraging behavior of ants searching for food.
ACO algorithms excel at solving combinatorial optimization problems, such as the traveling
salesman problem and the vehicle routing problem.

ARTIFICIAL BEE COLONY (ABC) ALGORITHM:

The artificial bee colony algorithm is inspired by the foraging behavior of honeybees.
In ABC, both employed and onlooker bees search for food sources (solutions) in the search
space, while scout bees explore new areas.

A.11   EJMD METHOD FOR AN UNBALANCED FULLY INTUITIONISTIC FUZZY SOLID TRANSPORTATION PROBLEM

In this appendix, we delve into the EJMD method as applied to an unbalanced fully intuitionistic
fuzzy solid transportation problem. This method offers a computational framework for addressing
complex transportation problems under uncertainty.

PROBLEM OVERVIEW:

The unbalanced fully intuitionistic fuzzy solid transportation problem involves the transportation
of goods from multiple sources to multiple destinations, where the demand and supply quantities,
transportation costs, and other parameters are characterized by fully intuitionistic fuzzy sets.
Unbalanced scenarios arise when the total supply does not match the total demand, necessitating
adjustments in the optimization process.

SOLUTION PROCEDURE:

Initialization: The EJMD method initializes a population of candidate solutions, representing
potential transportation plans.
Evolutionary operations: During each iteration, individuals undergo mutation, crossover, and
migration operations to generate new candidate solutions.
Fitness evaluation: Candidate solutions are evaluated based on objective functions considering
transportation costs, satisfaction levels, and other relevant criteria.

APPLICATION TO TRANSPORTATION PROBLEM:

In the context of the unbalanced fully intuitionistic fuzzy solid transportation problem, the EJMD
method adapts to handle uncertainty represented by fully intuitionistic fuzzy sets.
The algorithm aims to optimize transportation plans minimizing total costs while meeting demand and supply constraints under uncertainty.

ADVANTAGES AND CHALLENGES:

The EJMD method efficiently handles complex, nonlinear, and uncertain optimization problems.
Challenges may include parameter tuning, convergence behavior, and scalability for large-scale
transportation problems.

A.12 DISCRETE FOURIER SERIES USING GENERALIZED DIFFERENCE OPERATOR

In this appendix, we delve deeper into the concept of computing discrete Fourier series (DFS) utilizing the generalized difference operator. This method offers an alternative approach to analyzing discrete-time signals or functions, providing additional insights and resources for readers interested in understanding this technique further.

KEY CONCEPTS AND FORMULATION:

The formulation involves expressing the discrete signal or function as a difference equation using the generalized difference operator.

By applying the discrete Fourier transform (DFT) to the difference equation, one can directly obtain the Fourier series coefficients in terms of the difference operator.

ADVANTAGES AND APPLICATIONS:

Efficiency: Utilizing the generalized difference operator offers a concise and efficient method for computing Fourier series coefficients, particularly for signals represented by difference equations.

Flexibility: This approach provides flexibility in handling various types of discrete signals and functions, making it suitable for diverse applications in digital signal processing and communications.

Insights: Understanding the relationship between difference equations and Fourier series coefficients enhances insights into signal processing and mathematical analysis.

FURTHER EXPLORATION AND RESOURCES:

Advanced techniques: Explore advanced techniques and algorithms for computing Fourier series coefficients using generalized difference operators, such as fast Fourier transform (FFT) algorithms tailored for difference equations.

Research papers: Investigate research papers and academic literature discussing the application of generalized difference operators in signal processing and discrete mathematics.

Online resources: Access online tutorials, courses, and forums dedicated to digital signal processing and Fourier analysis to deepen understanding and practical implementation.

SOFTWARE TOOLS AND IMPLEMENTATION:

Experiment with computational tools such as MATLAB, Python's NumPy and SciPy libraries, and Octave to implement algorithms for computing DFS using generalized difference operators. Leverage built-in functions and libraries specifically designed for discrete Fourier analysis to streamline implementation and experimentation.

This comprehensive appendix provides additional information on computing discrete Fourier series using the generalized difference operator, offering insights into its advantages, applications, further exploration avenues, and practical implementation using software tools. Readers are encouraged to explore the topic further to gain a deeper understanding of this important technique in digital signal processing and mathematical analysis.

Glossary

Equation of heat flow: The equation for the flow of heat in time-dependent form is

$$D^2(\nabla^2\phi) = \frac{\partial\phi}{\partial t}$$

Here, D^2 is called diffusivity and is constant for a given medium. ϕ is the concentration of the diffusing material. The equation is used in modeling the heat flow through the medium.

Laplace's equation: This is a special case of Poisson's equation. The equation is

$$\nabla^2\phi = 0$$

Here, ϕ is scalar potential which is the same as the one in Poisson's equation. The equation can be used to find electric potential inside a uniform dielectric, gravitational potential in region of no mass, electric potential in charge free space, magnetic potential in current free space, etc.

multiple working vacation policy: In situation if this threshold is not met, the servers grab another working vacation and keep going with doing the same till the count of failed machines is N_1 or above toward the completion of a vacation period. This has been described as the *multiple working vacation policy* non-random variation.

Poisson's equation: The differential equation is

$$\nabla^2\phi = \rho$$

Here, ρ is called source density.

Queuing model: A *Queuing Model* is a feasible framework for expressing the service-focused problem in which customers arrive randomly in order to receive a service, with the service time being a random variable.

Schrodinger's wave equation: This equation is the fundamental equation of quantum mechanics. The equation in time-dependent form and time independent form, respectively, can be written as

$$-\frac{\hbar^2}{2m}\nabla^2\Psi = i\hbar\frac{\partial\Psi}{\partial t} - V\Psi$$

$$\nabla^2\Psi + \frac{2m}{\hbar^2}(E-V)\Psi = 0$$

E is the total energy of the particle, and V is the potential energy of the particle. Ψ is the wave function corresponding to the particle of mass m, and $\hbar = \frac{h}{2\pi}$ with 'h' is the Planck's constant. The equation finds an immense use in solving quantum mechanical systems.

Spanning soft trees:

Definition 1.1 *Let $H^* = (R,D)$ be a simple graph and $H = (H^*,W,Y,Z)$ be a soft graph of H^* which is also given by $\{J(z) : z \in Z\}$. Let $J(z)$ be a part of H for some $z \in Z$. Then, a subgraph $S(z)$ of $J(z)$ is called a spanning subgraph of $J(z)$ if the vertex set of $S(z)$ is the same as $W(z)$.*

Definition 1.2 *Let $J(z)$ be a part of a soft graph $H = (H^*,W,Y,Z)$ for some $z \in Z$. Then, a spanning tree of $J(z)$ is a spanning subgraph of $J(z)$ that is a tree. A spanning tree of $J(z)$ is denoted by $T(z)$ for all $z \in Z$.*

Definition 1.3 *Let $H = (H^*, W, Y, Z)$ be a soft graph of a simple graph $H^* = (R, D)$ represented by $\{J(z) : z \in Z\}$. Then, a spanning soft tree of H is given by $\{T(z) : z \in Z\}$ where $T(z)$ denotes a spanning tree of $J(z)$ for all $z \in Z$.*

Wave equation:

$$\nabla^2 y = \frac{1}{c^2} \frac{\partial^2 y}{\partial t^2}$$

$$\implies \nabla^2 y - \frac{1}{c^2} \frac{\partial^2 y}{\partial t^2} = 0 \quad \text{or} \quad \Box^2 y = 0$$

where $\Box^2 = \nabla^2 - \frac{1}{c^2} \frac{\partial^2 y}{\partial t^2}$ is called D'Alembertian. y can be the displacement of a stretched string, displacement of particle of medium as wave propagates, current in electrical transmission line, etc., depending on the nature of physical problem. c is the velocity of wave which is related to density of medium and its elasticity.

Index

Note: **Bold** page numbers refer to tables and *italic* page numbers refer to figures.